Books by Satish C. Bhatnagar

1. Mathematical Reflections, Volume I (2010), Volume II (2015)
2. Vectors in History, Volume I (2012), Volume II (2021)
3. Darts on History of Mathematics, Volume I (2014)
4. Epsilons and Deltas of Life, Volume I (2012), Volume II (2021)
5. Via Bhatinda, Volume I (2013), Volume II (2016)
6. My Hindu Faith & Periscope, Volume I (2012)
7. Plums, Peaches and Pears of Education, Volume I (2016)
8. Swami Deekshanand Saraswati: My Swami MamaJi (2014)

Publisher's notes

DARTS
ON
HISTORY OF MATHEMATICS

VOLUME II

SATISH C. BHATNAGAR

BALBOA.PRESS
A DIVISION OF HAY HOUSE

Balboa Press books may be ordered through booksellers or by contacting:

Balboa Press
A Division of Hay House
1663 Liberty Drive
Bloomington, IN 47403
www.balboapress.com
844-682-1282

Print information available on the last page.

ISBN: 979-8-7652-3559-1 (sc)
ISBN: 979-8-7652-3558-4 (hc)
ISBN: 979-8-7652-3630-7 (e)

Library of Congress Control Number: 2022919651

Balboa Press rev. date: 11/15/2022

READERS' COMMENTS

Visiting this museum is a part of my continuing approach of drilling into the nature of fuzzy mechanics of history as opposed to sharp linearity in mathematics.

In other words, any two mathematicians will either say that a proof/ solution is correct or wrong – no question of any split!..........if two historians completely agree with each other, then, it may be reasonably assumed that one of them is taking a free ride on the other.

Hey Satish, I see that you offer two courses in the History of Math at UNLV. We offer only the undergrad as of now. Would you be kind enough to send me the two syllabi for my reading and information that might be helpful if we someday offer a graduate course in the History of Mathematics here at Lipscomb University? I enjoy reading your letters, essays, etc.

Dear Sir, Many thanks for your email. I must say you possess an art of storytelling and your experiences in life plus everyday encounters with math problems continue to enlighten me.

I certainly agree with all of that. I truly hope unlv appreciates the real gem they have in you to be able to bridge the two disciplines. Have a great Thanksgiving!

I wish we had this class when I was a student. I've always enjoyed your lectures. You are definitely one of the best lecturers in my many math courses. You encouraged me to think, not just memorize.

Once again, I think you are spot on. You describe precisely what a HoM should be and what you make it out to be. Not sure anyone else should even be allowed to teach it. Well, except maybe me if/when you do retire! Thanks for sharing!

Satish, A very nice personal tribute and some more history of you. Thank you for sharing. So, you could have been a number theorist, but instead chose government service.

Dear Dr. Bhatnagar, Thank you for sharing. Slowly I am trying to include more culture in my instruction. My students are doing a project on Ethnomathematics and they are enthused to explore how other cultures valued mathematics.

DISTRIBUTION OF CONTENTS

II. HUMANISTIC SLICES

III. INDIA SPICES

IV. SMORGASBORD BITES

V. OTHER PERSPECTIVES

DEDICATION & FRONT COVER

Normally, I dedicate a book to a person who has inspired me to write it or to a person who has assisted me in an extraordinary manner. Sometimes, I dedicate a book to memorialize a deceased person who has left an indelible mark on my life. In the context of this book, there is no one person standing up in my mind.

Looking back to my dedication, **'History for the Wise'** to Volume I, I realized that it was not completely correct as Volume I and Volume II of ***Darts on History of Mathematics*** are precisely on the History of Mathematics - not just about History in general - though, my approach to the History of Mathematics is getting highly humanistic.

Anyway, instead of leaving this page blank, I thought this note would fill a void. Who knows in near future someone or something may emerge who is worthy of the dedication of Volume II, or Volume III, which I am not going to rule out. So far, it seems that volumes on **Darts** have been evolving from my reflective writings alone.

As far as the cover design of Volume II is concerned, it is isomorphic to the one that I did design for Volume I. This is natural as that also reflects the similarity of their contents. Essentially, there is a permutation of colors. At the same time, Volume II has to look different in some respects. The world map, the darts and arrows, my picture, my bio and readers' comments have the same placements and visual feel. However, I remain open to any change that may be suggested by the design department of the publisher.

Satish C. Bhatnagar
June 20, 2021

DARTS ARE VECTORS TOO!

(Modified from the Preface of ***Vectors in History, Volume I***)

The word 'Vector', in the title of the book, ***Vectors in History*** (2012) is borrowed from mathematics, where vectors are not restricted to only two or three space dimensions, as generally used in physics and applied mathematics. Vectors represent quantities that need both magnitude and direction for their representations. Visually, vectors are shown by directed arrows or darts 'freely flying around'.

Each reflection is like a vector or dart. Its magnitude is subjective, if measured by its impact factor. Vector is tangible, if it is measured by the number of words in it, varying from 600 to 6000. Its theme provides a direction; however, it may change its course during its flight. In other words, a particular reflection may have more than one attractor point. In mathematics, a zero vector has zero magnitude, but no direction. Naturally, it does not correspond to any reflection!

As far as the variety of directions of a vector is concerned, they are 'doubly' infinite in 2-dimensions and 'triply' infinite in 3-dimensions! These reflections do bring out a similar flavor, when it comes to the number of sub-topics that are touched upon. They are not bounded above - be that in the context of a country, person, religion, or region etc.

These reflections are truly like free vectors having any arbitrary initial and terminal points. They can fly off from anywhere and land anywhere. In the world of mathematics, two vectors are defined equal, when their magnitude and directions are the same. This is never a case in my reflections, as any two of them would differ both in words and themes! Besides, who would care to read 'equal' reflections?

In mathematics, two position vectors/forces can be added. In these reflections, there is generally no connection between the ending of one reflection with the beginning of the next. Consequently, one can read any reflection from anywhere without missing a beat from the previous one! Often there is more than one strand in a braid, and my style remains compact. Nonetheless, these reflections are snappy and penetrating - like darts and arrows - vectors!

A general structure of a reflection is that it spins off from a specific incident. I basically dive into a vortex with it, or look out through it as a window to the universe far and beyond. In the process, it is wrapped with other stories and concepts which are often cross-cultural and interdisciplinary. The delight they have given me on subsequent readings, I am sure, they would give in different measures to each reader. That is a kind of beauty and uniqueness of this book.

Satish C. Bhatnagar
Dec. 16, 2020

A HIST-O-MATH PREFACE (PART II)

A few words need to be said about how certain things happen in life without any planning. There is no preface with a title, ***A HIST-O-MATH PREFACE (PART I)***. Reason, there is no ***Darts on History of Mathematics, Volume I.*** When I finished the ***Darts on History of Mathematics*** in 2014, I did not foresee Volume II ever coming up. That is why its title did not have Volume I on it. ***Darts*** is a mathematical corollary of my book, ***Vectors in History***, which I never thought of writing.

In order to bring closure to this thread, until the age of 60, it was not even in my dreams to see my name on any book. That is the beauty of life! Both volumes are independent in the sense one does not have to read Volume I before embarking upon Volume II. Therefore, this preface is modified from the preface of 'Volume I'. Instead of putting the following texts in a series of quotes and italics, I am just adding an extra line space.

Writing the preface of a book is a crowning moment in an intellectual endeavor. The important things about a preface are the whys, hows, and whats about the book and its author. That is what follows:

How much mathematics is needed to understand this book? A little more than high school level! A history of math course can be taught with little math or with good doses of math - depending upon students' background. No one is interested in the history of any one mathematics problem or a topic - it is boring. Yes, survey papers are possible. What is really needed is a mathematically mature and curious mind which can also navigate the choppy and muddy waters of math and history.

What are my credentials for writing this book? My inclination towards the History of Math (HoM) has been organic. In the 1980s, I started cutting away the umbilical cord that connected me with my PhD research. I was then groping around and trying to discover my new strengths. It is a thesis that before the age of 40, there is no meeting ground between one's knowledge of mathematics and history - the two remain poles apart. Around the age of 50, one begins to sense a confluence of deductive thinking of mathematics and soft thinking of history. It started gaining traction in my mind.

The ***Darts on History of Mathematics*** is a corollary of my book, ***Vectors in History, Volume I*** (2012). Before writing a book on the HoM, I needed to establish my identity as an historian, as I approach history of mathematics as a subset of history in general. Also, if you browse standard textbooks on the HoM, they are stereotypical and isomorphic. The contents of early history are perpetuated from one author to the other without questioning. After all, who has the time and motivation for 'revisionism'? The books are written by math professors who seldom developed a sense of history.

Long bibliographies and lists of references at the end of the books, and sometimes notes at the end of each chapter, overwhelm the readers. The facts go unchallenged - particularly, if they suit a particular academic culture. Years ago, when I questioned Morris Kline on mathematics in ancient India as written in his book, ***Mathematical Thought from the Ancient to Modern Times*** (1972), he wrote me back saying that he had only quoted from a book by Indian authors. Can a wrong proof of a theorem or a wrong solution of a problem ever be passed on by such referencing in mathematics? Never! Over the years, these padded bibliographies and references have lost my respect. They are not a part of any one of my books.

What prompted the writing of this book? The darts in the title of the book precisely point out towards such unverified, exaggerated and stretched out or ignored conclusions in their historical narratives. Such scenarios urge me to write these reflections on specific points. Therefore, ***Darts*** is not a typical textbook on the HoM, but it can be used as supplementary material along with a 'regular' textbook.

What is the incubation period of the book? The dates of some reflections do go back to more than 20 years. However, the writing of a book was not on the horizon until relatively recent. Also, not all old pieces of my writings were saved.

What is new in the book? Apart from its format; in brief, it has thought provoking angles of observation and deductive conclusions on many topics, which may look ordinary or rare.

Who will benefit from the book? Any lay person with an historical bent of mind on mathematical topics stands to gain from it. Both undergraduate and graduate students in history of mathematics courses would enjoy it. All reflections are independent - they are excellent bedtime reading too, which is never suggested in the learning of hardcore mathematics.

Is there any other book similar to this in format and style? Absolutely none! That is my moment of pride. It aligns with my disinclination for writing mathematics textbooks throughout my professional life of 50+ years. I do not see any intellectual challenge in the writing of one. It is uncreative to shuffle problems from other textbooks and change them by epsilons and deltas! A computer may be doing it.

Volume I has 71 reflections and write-ups in this book, but Volume II has 78 of them. They are unevenly divided into the same five sections. The divisions are not sharp as dictated by the very nature of reflections. Sometimes, the overlap between topics in some reflections is so large that their variants are also included in the ***Converging Matheriticles*** (2014) or/and ***Vectors in History*** (2021). I have realized that some divisions of reflections are better than no division, as it was done in my first book.

Reflections under the heading of **Classroom Cuts** are drawn from classrooms. The second section, named **Humanistic Slices,** contains reflections connected with the lives of mathematicians of both past and present. The third section, called **Indian Spices**, has assorted reflections dealing with mathematics of India - from ancient to the present. The fourth section is called, **Smorgasbord Bites.** As the title implies, it has an interesting mix of reflections. The fifth section includes the outlooks on the HoM of my friends, colleagues, and students of mathematics. It is rightly called the **Other Perspectives**.

A common feature of all my books is that they each can be read from anywhere, as the reflections are independent in contents. It fits in today's fast life styles; no one has the time to start a book from its Page Number 1 and then wade through it to the very end. Brevity is the characteristic of the twitter age - using 140 characters or less. As a consequence, abbreviations are explained again, and certain references repeated as encountered.

The practice of dating each reflection is continued so that a reader may have a full perspective of its genesis in terms of time, place and my mindset. As a young reader, I paid no attention to it, but now, this is the first thing I look for in a book. The two dates on some reflections means that a significant revision was done on the second date. Another continuing feature is providing partial and full blank pages for the readers to scribble their comments, as they pop up while reading it. It comes from my compulsive habit of underlining and side-lining a significant part of a sentence or paragraph. Such markings become a source of quick reference in future.

Every book is a defining moment in the author's life. For me, this Volume II firmly plants my feet in history. Increasingly, I look at a socio-political scenario through a lens of history. Consequently, my world of solid mathematics has shrunk. I have not volunteered to teach hard core graduate courses for nearly 20 years. Even my bread-and-butter courses on ODE have distanced from me.

The main reason for a change in my teaching load is due to the creation of ***Teaching Concentration*** (2002) in the MS program. My involvement has increased in the teaching of its three required courses. There are a few faculty members who can teach the hard-core math courses that I used to teach, but lately, I have been the only faculty member teaching these specialized courses in this concentration.

Acknowledgements. For writing Volume I, I got sabbatical leave for Spring-2014, and for writing Volume II, I got another sabbatical leave for Fall-2020. I thank UNLV and my department chairmen for their support. Also, I profusely thank Scotsman, Francis A. Andrew, a science fiction author and Professor of English Language (working in Oman for the last 15 years), for providing me feedback on the syntax and semantics on every reflection that I have written since 2009. With the result, along with my unabated passion for writing, I am beginning to enjoy the usage of the English language in a manner never ever done or thought before! Finally, any comments and suggestions on the book emailed at viabti1968@gmail.com, would be greatly appreciated.

Satish C. Bhatnagar
June 22, 2021

GLOSSARY AND ABBREVIATIONS

Mantra is a set of supposedly energized syllables in Sanskrit – potent enough to affect material changes with right repetition and enunciation.

Sutra is a cryptic and condensed description of a principle or property. An example is of 18 *sutras* of Vedic Mathematics.

Tapa is combination of penance, meditation with austerities

Vedas refer to the most ancient four Hindu scriptures, namely; Rig, Yajur, Atharva, and Saam. ***Upvedas*** and ***Vedangas*** are ancillary treatises for a systematic study of the Vedas.

Rishi is an enlightened individual in terms of his/her cultivated powers of mind developed through Yoga over a long period of time.

Guru is far more than a high school and college instructor. There is an associated element of one-one-ness, loftiness and holistic nature - bordering spirituality.

Gurukul (a kind of Hindu seminary school going back to the Vedic period)

Shrimad Bhagwatam or Bhagwat is a holy scripture. It is often confused with Gita or Bhagwat Gita, which is a central part of a chapter in the epic of Mahabharata. Shrimad Bhagvat was compiled by Vyas after he had finished the Mahabharata. It is a great story of life.

advaita; non-duality - Principle of one-ness.

siddhis are the states of mind achieved after years of penance and yoga that one can materialize objects. Essentially, it is a reverse mass-energy equation. The late Satya Sai Baba of Puttaparthi had a reputation for pulling out jewelry items of thin air for his followers all over the world. Such a person is called ***Siddha.***

IT: Information Technology

BTI: Bhatinda or Bathinda

DLH: for Delhi, the capital of India

UNLV: University of Nevada Las Vegas

JMM: Joint Mathematics Meetings

MAA: Mathematical Association of America

AMS: American Mathematical Society

NAM: National Association of Mathematicians (Founded by Afro-Americans)

AWM: Association of Women in Mathematics

SIAM: Society of Industrial and Applied Mathematics

IMS: Indian Mathematical Society

Math: Popular abbreviation for Mathematics - used interchangeably in the book

PDE: Partial Differential Equation(s)

ODE: Ordinary Differential Equation(s)

SECTION I

CLASSROOM CUTS

1. MINING MATHEMATICS IN MUSEUMS

(A Note to my students in the History of Mathematics course)

Our next class on Tuesday, 9/24/ 2013 will be held inside UNLV's Barrick Museum - just east of the Lied Library. I have already spoken with a person-in-charge about it. Please be there before 4 PM, if you can, as the Museum closes at 5 PM.

Visiting this museum is a part of my continuing approach of drilling into the nature of fuzzy mechanics of history as opposed to sharp linearity in mathematics. You must be able to form some ideas as to how facts in history are created, communicated, perpetuated, constantly revised, and even challenged.

Before the museum visit, I would like you to review the **first three chapters** of the textbook – wherein, particularly various artifacts are over-emphasized for their historic values for mathematics.

You can get a lot out of this exercise. Observe the museum objects in silence – refraining from discussing anything with each other. Take individual notes. You will then understand subjectivity in the facts of history.

You are free to stand in front of any collection of artifacts for any length of time. In other words, you don't have to check out every showcase in the Museum, or of any museum in the world for the same reason!

It is better if you don't read the entire descriptions of the exhibits. However, after you have extracted and written out any mathematical or scientific information, then you may come back to it at your convenience, read the descriptions and draw some comparative conclusions.

I do not undermine the significance of any 'ancient' objects, but only caution you about drawing too many conclusions. We all know how human DNAs are loaded with information, but extracting that information requires incredibly sophisticated chemical analysis run in high-tech labs. So, you should scrutinize any mathematical findings in any textbook from this angle.

My suggestion on working during this museum visit is that you enter only one main 'thing' on one page of your notebook. Briefly scribble the ideas generated from visual examination. You are not touching any object, tasting it, smelling it, or hearing it! Your angle of vision is limited to 120 degrees.

You may check at the Museum office, if taking pictures of the items is allowed. In summary, approximate the age of an object(s), compare it with anything in the present. Derive mathematical awareness and knowledge of that society in terms of its geographical extent.

A report on mathematical findings will be your exclusive next weekly project. So, mull over this visit for a few days before putting down your thoughts. In case, you need to go and reexamine any exhibit, please do so out of the class hours.

Finally, I want this experience embedded into your psyche so that any time you visit any museum anywhere, you would examine its displays with as much open mind as possible. Your sharing any mathematical nugget with me will be appreciated.

Sep. 21, 2013

COMMENTS

Hi Satish, Your reflection reminded me of a book entitled On Stonehenge by Sir Fred Hoyle. This book had a large contingent of a discipline which Hoyle had devised - namely "archeoastronomy." Hole had demonstrated in this book that Stonehenge was in fact an astronomical observatory designed mainly for the prediction of solar eclipses. Hoyle argued that conventional archeology was essentially funerary in nature and generally involved the retrieval of the remains of artifacts from the remote past. By showing how Stonehenge was designed as an astronomical observatory, Hoyle contended that the past could be brought to life. One could imagine the festivities associated with the astronomical calendar, the colourful costumes of the dancers, the melodious chanting of the singers and the general merrymaking of those attending these joyful performances.

Perhaps, you could extend Hoyle's discipline of archeoastronomy to embrace archeomathematics. By observing the geometrical, trigonometrical etc. abilities of ancient artifacts, one could deduce the mathematical competence of these ancient peoples and so come to conclusions about their technical, architectural and artistic achievements. One could then determine how their mathematical abilities influenced their religious and philosophical systems. All best: **Francis**

Satish, very interesting. Is this the first time you are doing something of this sort? **Noel**

2. SWITCHING THE HATS

Teaching a course on history of mathematics, particularly to math graduate students, presents a unique challenge. By the way, UNLV has two courses on the history of mathematics. Math 314 is at the undergraduate level and is required for the secondary education math majors. The second one, Mat 714, at the graduate level, is required for MS in the Teaching Concentration - one of the four MS tracks. Both the courses are offered every two years.

A challenge primarily lies in changing the mindset of the students in their learning about historical events. As a matter of fact, it applies to the instructors too. One may say - what kind of mindset is called for? It is simple to explain. A graduate student in mathematics has normally taken over a dozen solid mathematics courses after getting into a college. S/he (means she/he) fully understands what a proof means in mathematics; clearly knows the difference between a proof of a statement in mathematics and its verification, and also appreciates how a single counter example can dis-prove, rather demolish a statement into falsehood.

A corollary of the above characteristics of mathematics is that there is no disagreement between mathematicians once a proof of a theorem is settled. No one cares, if it is proved again and again – unlike, say, the playing of Mozart's symphony in the world of music! However, there may be different 'stylish' proofs of a proposition or different solutions of a problem, but they are all logically consistent with the previous mathematical results. In other words, any two mathematicians will either say that a proof/solution is correct or wrong – no question of any split! A mathematical proof is not based upon individual opinions, but rather on absolute understanding of a collective body of mathematical knowledge of an area.

In the realm of history, the story of facts, which essentially correspond to statements in mathematics, is very different. Putting the issue right away on an anvil, when it comes to researching, rehashing, or reporting a historical event, if two historians completely agree with each other, then, it may be reasonably assumed that one of them is taking a free ride on the other. The 50th anniversary of President Kennedy's assassination is a perfect example.

So many books have been written, thousands of expert interviews have been recorded on the assassination, but the whole area is 'muddier' and muddled than ever before. Interestingly enough, new conspiracy theories keep showing up and rumor mills have not stopped churning them out.

I use one textbook for a typical math course as well as for a history of math course, but their functions are very different. Math textbooks are rarely used to explain the proof of a theorem, or an example already worked out. Its utility is for assigning homework, discussion and dissection of a couple of typical problems. In a math course, it is enough, if the students are able to understand the material covered in the previous meeting. Rarely, students have time to look at the new material ahead of its class presentation, though this practice is gently recommended.

This scenario is different in a history course – at least in the way I approach it. Here, it is essential that a critical reading of some material is done ahead. Of course, the students do the homework based upon the material covered in the previous week. That is another feature of my history course. During lectures, every time, students are invited to comment on one or two salient points or make 5-minute presentations on the material studied, it invariably ends up in the gist or summary. There are no questions and strong disagreements with the author or observations therein. In order to instill the difference between mathematical and historical facts, from Day Number One, I tell the students to take off their math hats and wear different hats of history as soon as they enter the classroom.

As for myself, I find disagreement and new angles of observation penciled all over the pages, while reading the same material. My thrust is to literally leave off the pages of the book and see them as a part of a bigger picture in terms of society, nation, political systems, organized religions etc. in order to understand why a particular math problem was handled there or a body of mathematics was done, and so on.

For the American students, a textbook is a must, as their faith in it runs very deep. Consequently, they do not develop good note taking skills. Also, most UNLV students having time jobs and/or having families, they do not have relatively enough study time outside the class. During my college days - through my bachelor's and master's, I was a full time student – meaning no part time jobs, which neither existed in India of the 1950s, nor were we encouraged due to the brutal pressure of the annual comprehensive examinations.

I am sure my students will remember this course for the class discussions whenever they would study anything from historical perspectives. At times, during textbook readings, I would literally fly away 'too far' to lose connections with the textbook material. It comes out of my love - hate – love type of relationship for history since the age of 15. I never kept a long distance from history. Naturally, history has seeped into my system as deep as mathematical thinking has. I am a rare breed intellectually.

Nov. 27, 2013

COMMENTS

Dear Professor Bhatnagar, I loved your writing on teaching the History of Mathematics. You have perfectly brought out the fact that students (and teachers, if I may say so) have to switch their hats when they discuss the history of Mathematics as opposed to discussing Mathematics itself. Your contrasting Math and Music is very apt. No one would like to read a proof of Rolle's theorem every time he sees it but he would like to listen to a melodious film song every time it is played on the radio. This means that a well composed musical piece is never transplanted by another but an elegant proof of a theorem is soon relegated to the backwaters of Mathematics by other "more" elegant theorems. Thanks for a thought-provoking piece of writing. **Arun Vaidya**.

Dr. B, I've never had a course in history of math, but I've picked up quite a bit here and there out of an interest that I trace back to a History of Chemistry course that I took at UW-Madison in 1959. It was a 1-credit course taught by a senior faculty member, and a welcome relief from the traditional history courses where the emphasis is on political events, wars, treaties, dates, etc. **ONN**

Yes, there is a lot of regurgitation of the history of popular topics, that don't add much that is new, because they sell. However, historians revisit subjects because they raise new questions that were ignored previously. Good historians also stress the significance of events in light of what transpired subsequently. That is why we have different interpretations. New voices are heard. So women were ignored for many centuries, what they thought and did. But to figure out why people made certain decisions one must judge things on the basis of hindsight. Do you get your students to discuss the significance or importance of a theorem in light of the appearance of new theorems? Be well. **Noel**

3. FOCI ON HISTORY OF MATHEMATICS COURSE

The 2013-fall semester will be over after a week.The ***History of Mathematics*** (MAT 714) is the only course that I look back at in terms of how it has evolved so far, as it provides me with the most enriching experience and thoughtful moments. It is definitely helped by an engaging group of students that I always have. Out of the ten students enrolled this semester, four are full time graduate assistants, three high school teachers and three full time academic employees at UNLV. The textbook used is ***A History of Mathematics*** by Merzbach and Boyer, 3rd edition (2011), It may be adopted one more time for a couple of reasons.

Since 2007, this is my fourth offering of this course. There is a progressive process involved both in my growing sense of history and what I want to convey to the students with my stamps on certain topics. It does not happen in a typical math course despite my delivery of those courses being offbeat. It is the nature of mathematics which does not have any malleability in its instruction. For my own knowledge and check-ups, here is a list of markers that have been, by and large, touched upon during this course. The numbering does not signify any order – it is the way they came off the top of my head.

1. The main thrust on any history, going back to over 2000 years, is not to lament over the unavailability of books in the modern sense of their availability both online and offline. Ancient history has to be extricated from any ancient objects and surviving monuments - like the pyramids of Egypt and other structures. They are the real DNAs of history and we do have scientific tools to unravel them. For instance, it is often reported - how certain crimes, unsolved for decades, are resolved with the help of new forensic knowledge that did not exist earlier.

2. A degree of caution has to be exercised whenever sweeping conclusions are drawn from archeological findings taken out from a millionth portion of a land, very small as compared with the rest of the land. Archeological work is very expensive and any chunk of research money always comes with an agenda. Therefore, I urge my students to be intellectual vigilantes

while studying history based on anthropology, geoscience, archeology, sociology, etc.

3. The ancient works, both mathematical and non-mathematical, associated with famous names like those of Pythagoras and Euclid, are partly attributable to the groups of researchers clustered around them or/and academies that they must have nurtured as founding heads.

4. Any aspect of the history of mathematics is better understood, if it is viewed as a subset of contemporary political and social history.

5. Both mathematics and modern science flourish in organized societies. Also, rapid progress in science and mathematics is about the same during times of war and peace - specifically, it is applied in war and pure in peace.

6. Science, mathematics and engineering flourish during the building phase of a nation or civilization and the progress slows down during its declining period, when ethics, arts and humanities rise up in public prominence – fitting in the context of rise and fall of civilizations.

7. In no age and place, mathematics is distributed uniformly. There are mathematically dark spots in every nation in every age.

8. Since the early eras to the modern ones in all cultures, the relative presence of women in mathematics has been small. However, based on my data and experience of the last 50+ years, it is due to women's priorities in life rather than due to any gender inability to do fundamental research.

9. The colonizing nations - England, France, Spain, Italy, Portugal and Germany - were at the same time also leading nations in mathematics and sciences too.

10. There is a statistical correlation between mathematics and organized religions. It is one of my ongoing projects to identify mathematicians, Nobel Laureates, and scientists from major religions; namely, Judaism Christianity, Islam and Hinduism.

11. New math, produced in the universities, by gifted researchers in conducive environments is provided by good administrators, especially chairmen and deans. The role of department chairs in the development of a good department cannot be overemphasized. Rarely a world class mathematician is a good teacher or leader of a department. On the other hand, not-so-great research mathematicians have built up great mathematics departments.

12. New branches of mathematics are constantly being created by questioning the foundations and fundamentals of old systems. Geometries provide perfect scenarios.

13. As a part of colonization, the native cultures were historically trivialized in terms of their languages, literatures and mathematical awareness. It is evident from the decimation of Mayan, Aztec and Inca civilizations of Latin America by the Spaniards, Balinese by the Portuguese, Haitians by the French, and South Africans by the English. It is the price native people pay for not being politically awake and united. In this respect, the Hindus, despite being known for their intellectual achievements and spirituality, lost Hindustan/India, their only homeland, to successive invaders from the Middle East, Central Asia, and Western Europe.

14. Progress in science and mathematics is neither strictly linear (also called Darwinian), as commonly projected, nor a two dimensional aperiodic function. It is closer to a three dimensional sinusoidal surface. It is based on the fact that the phenomenon of permanent loss of a household item by an individual or family, can happen to an entire society, nation, or civilization too. Thus are created dark periods or black holes in history.

15. Trivialization of the past and its history is simply based on the paucity of ancient facts.

While concluding, itt does not mean this is over - a few more points and themes are not cropping up in my mind. At this moment, call it mental fatigue or memory lapse! It is time to close it out for the time being, and get help from outside. After all, mathematics and history are collaborative disciplines too!

Nov. 30, 2013

COMMENTS

Hey Satish, I see that you offer two courses in the History of Math at UNLV. We offer only the undergrad as of now. Would you be kind enough to send me the two syllabi for my reading and information that might be helpful if we someday offer a graduate course in the History of Mathematics here at Lipscomb University? I enjoy reading your letters, essays, etc. Remember we were together in Peru a few years ago (2007 I believe). We were roommates in a few places. Sincerely, **Doy O. Hollman**

Well said! I really like your point that "any chunk of money comes with an agenda"...can also be true in educational research as well. **Aaron**

I wrote: I think you were in my first offering of Mat 714. Tell me 1-2 things that stand out after 6-7 years.

Aaron: I think a couple things stand out--we did an interview for the 50th anniversary of some department members that had been with UNLV for many years. It was interesting to get their insight. Probably as no surprise--I actually liked your reflections as well. They are very thought provoking and it's also nice to see that your instructor is also a student/academic (but with much more experience and content-knowledge). Can't remember if it was that class or another, but I believe we did some sort of conference at the library that was helpful. Also really liked the textbook myself. Happy Holidays!

Aaron: Oh, and how could I forget--my signed copy of the complex variables book with a nice personal note with it! I believe students tend to work harder in courses when they know the Professor knows them by name and cares that they continue to develop intellectually throughout the course. I've never had a better math professor in that realm than you. Cheers.

Satish, Very interesting report. You make excellent points. Do you have any ideas of why math is connected to organized religion? I like your term "intellectual vigilantes." Why do you think undergraduates avoid the course, especially those planning to teach math? Best wishes, **Noel**

Professor Bhatnagar, Let me propose a possible #15 for your list. You've said this more than once to include during yesterday's class:

15. Remnants of a civilization's/society's math and science achievements are more easily discarded over time than that of art and literature. For lack of comparison, I would be hard pressed to cite good examples of this. Though in chapter 6 at page 125 (last paragraph) we see where Archimedes works were written over. Respectfully, **Bob Lynn**

Jai Siya Raam Prof. Bhatnagar -As I said before your write up is a 'good read'. You have invited comments, here are a couple:

1. In your point #1.... "Ancient history has to be extricated from surviving monuments", I would add .... monuments, social and religious traditions and rituals". An example is that the concept of ZERO is preserved in our daily ritualistic prayer "***Poorn-madah poornm idam*** ...." This ***Sloka*** is over 2000 years old. I did notice in your subsequent points mentioned religious and social. aspects.

2. Your statement # 10. "There is a statistical correlation between mathematics and organized religions. It is one of my ongoing projects to identify mathematicians and scientists from major religions of Judaism Christianity, Islam and Hinduism." is in agreement with me. My favourite example here is the concept of the (so called) Binary system as it existed in time of Pingali/Panini (2800 BC) and is well preserved in our prayers even today. Cheers - **Hansraj**

4. A PITCH FOR HISTORY OF MATHEMATICS

During the last 41 years, at UNLV, I have never made a pitch to the students for registering into my courses, especially those, which may be canceled for the lack of low enrollment. Well, as goes a saying - there is always a first time for anything in life. Or, it is one of those cryptic Murphy's Laws: ***Whatever may happen will happen one day***.

During this Fall - 2015 semester, I am scheduled to teach a graduate course on ***History of Mathematics*** (MAT 714). As of today, a week before the semester is to start, seven students have registered. This being a course on History of Mathematics (HoM), a line or two on its 'history of enrollment' is also in order! Since Spring-2007, besides other instructors, I alone have taught this course four times. The first day enrollment numbers have varied from five to ten. So, 'lucky seven' makes a safe bet for it to run, as the numbers in most 700-level math courses are under seven. However, here are some of the following reasons to keep this course in mind for now or future:

1. History of Mathematics is a very important course for the undergraduate as well as for the graduate students. It is immaterial how and what in life one is going to do with mathematics, one would be better off knowing its history. No history department will offer this course, as their faculty members rarely take mathematics beyond Precalculus II. The US college curricula justifiably boast courses on the history of any topic under the sun - from Rock and Roll music, to philosophy, to sciences, to any aspect of urban life etc. It is a reflection of the greatness of the US as a nation too.

2. The graduate course, MAT 714, is offered every two years. The main reason: it is required only in the ***Teaching Concentration*** - one of the four concentrations for an MS in Mathematics. Any graduate student not taking this course now will miss it at UNLV for a long period of two years.

3. In a graduate program, there is always room for one elective course. The dividends of this course, over a long haul, are immense. On a personal note, as a doctoral student at Indiana University, Bloomington, I took nearly a dozen elective courses in all areas of mathematics except mathematical

logic! These electives have enriched my professional life to the extent that I have taught nearly 75 different regular and experimental courses including honors seminars at UNLV alone. Furthermore, eight (to-date) of my books are published in six different genres - a rare feature for any author!

4. An HoM course is seldom taught by an instructor who has also studied general history. Yours Truly has passed a 10,000-hour benchmark on general history. Besides, a solo-authored book on history, called ***Vectors in History*** (2012) has been written. If anyone can find any other author of HoM book(s) with background both in mathematics and history would be taken for an Indian lunch buffet. Google or ogle for this information in the Lied library.

5. My approach toward HoM is holistic, as it is viewed as a subset of history in general. Thus, the ***Darts on History of Mathematics*** is a corollary to ***Vectors in Histor***y. The ***'Darts'*** is not a typical textbook, but a unique collection of reflective articles for supplemental reading to any traditional textbook on HoM.

6. During the course of study, the emphasis is laid out on the concepts of facts, verification, and proofs in both mathematics and history. A hands-on math history project makes a lasting impression. Moreover, a course on HoM sheds light on a question if mathematics was discovered or invented. The diverse backgrounds of the students in this course invariably enlivens the class discussions.

7. During the April-2015 Sectional Meeting of the American Mathematical Society, Yours Truly organized the first ever session on the **Necessary and Sufficient Conditions** for the development and flourish of mathematics at various levels - from that of an individual to a nation. It is likely to be repeated at a joint national meeting.

I am stopping at **Number 7** for obvious reasons. If you still have any question, curiosity, or concern about this course, please call or send me an e-mail.

August 16, 2015

COMMENTS

Dear Satish-ji: Thanks for sharing your blog on the history of Mathematics and your efforts in promoting its teaching. Apparently, you have done yeoman work in this field, not to mention other areas. I know that but for the invention of Zero by Aryabhatta (Disclaimer: not a relative!) both mathematics and science would have remained in a primitive state, to this day. As a youngster, I recall reading the book, "Mathematics for the Million " by Lancelot Hogben, who did acknowledge India's contribution of Zero. I believe this was no mean achievement, although the numerals invented in India were called Hindu-Arabic numerals, as the Arab traders brought them to the West. Thanks for letting me recite my tiny piece. With regards.
Mohan Bhatt

5. FIRST ASSIGNMENT!

"Every person has two biological parents, four grandparents, eight great grandparents, and so on, to any power of two. I want you to find out the name of any one of your eight great grandparents. Also, in about 100 words, include any mathematical tidbit of his/her time - be that from a family yarn, school, college, or elsewhere. Resorting to oral traditions is absolutely encouraged."

This has been the first assignment on the very first day in a ***History of Mathematics*** (MAT 714 and 314) course taught lately both at the graduate and undergraduate levels. Initially, it tickles the students, as most of the students being math majors, are conditioned to expecting raw mathematics in courses with MAT prefixes. But this is not a typical mathematics course, per se. It may be called a restricted history course, like the concept of ***Restriction*** of a ***function*** in mathematics, which is defined on a subdomain.

What is the purpose of this assignment?

This question can be posed at any genealogical level. The grandparents exist at every level, but the difficulty of finding relevant information about them grows too. This scenario is a perfect example of a cliché - that it is easier to pose a question than to find an answer. The Mighty Time, as the Ultimate Destroyer, has a tremendous impact on general history, especially on the History of Mathematics (HoM). Therefore, digging out facts from the past becomes exponentially challenging, as one goes back in time.

Another purpose of this exercise is to impress upon students the nature of facts in history, and how they are different from the ones in mathematics. There are no emotions associated with the facts in mathematics - whether they are encountered in axioms, definitions, lemmas, or theorems.

Also, this course impresses on the students that for finding facts in history, you cannot simply Google them, or expect to hit them in some books in a library. In HoM, the footwork or legwork is also called for. That means some cultivated skill in communication is desirable, particularly, when it comes to interviewing individuals for information - called oral history - a legitimate mode of history.

Interestingly enough, this simple question becomes absurd or funny when it is projected into the future in terms of one's children and grandchildren. However, in mathematics, such a sidebar exercise often yields newer insights. It is partly due to the differences in the nature of two disciplines. History is outwardly more human-centered than mathematics is.

Students are given a week to bring back their findings and share them in the class. That becomes fun and educational for everyone. I always look forward to reading them out. I have not kept any statistics on it, but hardly 25 % of the students are able to find out even the names of their great grandparents. Human alienation is bound to increase, as the societal radius shrinks in the fast digital lifestyle.

August 22, 2015

COMMENTS

Dear Bhaisahib: It is a very interesting assignment. I don't even know my grandparents' names. I wonder how the new generation these days can answer when their parents are divorced and they have almost no connection with their grandparents. If you say 25% can answer it is a large percentage. Thanks for making me part of your essays. **Girija**

A very clever and intriguing assignment to your youngsters. regds **jbp**

6. GLANCING AT DISCRETE MATHEMATICS

For several reasons, a course like ***Discrete Mathematics*** (Math 251) is unique in the undergraduate curriculum. Giving a touch of history, this is the last standard course, at a lower division level, which was created in the 1980s! The main reason was to bring back the 'proofs' in the curriculum. Since the 1960s, the proofs were gradually thrown out of the math textbooks due to the pressure from the engineering departments, who thought proofs in math courses for the engineering students were over emphasized at the cost of applications. I remember a calculus textbook published in the 1980s by the elite engineering organization IEEE that did not have even a single statement of a theorem!

In life, sooner or later, there is a reaction to an extreme stand. By the1980s when the discipline of computer science matured a bit, the students needed mathematical proofs for their courses in algorithm analysis. At UNLV, this course was created at the demand from the computer science faculty. Immediately, it also became a required course for the math majors and minors.

Before the 1960s, the proofs, which were embedded in precalculus and calculus courses, are now reincarnated into one course. It has several course names. On a personal note, my exposure to proofs was through Euclidean Geometry introduced in the 10th grade of high school in India. In the British colonial system of education, proofs were integrated in every math course.

The topics in this course are not new. I tell my students on Day Number One that look at the titles of the first four chapters that are to be covered, and browse the problems in the exercise sets. There is at least one problem in each exercise set that most students would recognize. For example, ***the sum of two even integers is even,*** has been 'known' since high school days.

What is new then? It is the proof of such statements. A course on discrete mathematics is all about proving statements, verifying statements, but emphasizing at each opportunity that verification is not a proof. Finding counterexamples to disprove some statements is a kind of brain twister.

When someone provides a proof of a theorem, then he essentially answers every question that could be raised. That is why in mathematics alone, the proofs of problems are not decided by a show of hands. Never!

My approach is to take advantage of students' familiarity with some topics and problems, and demystify the whole enchilada of proofs. The technical notations are minimized, as they hinder the writing of the proofs at this level. Deductive thinking is stressed - that is giving a reason for every step taken.

This textbook was not written for the UNLV students - no author does it for one school. The authors want their textbooks sold to all schools. Incidentally, American textbooks have the largest overseas market. Therefore, my jumping around the sections is my way of customizing the material for my students. In order to catch a wayside nugget, students are encouraged to read the textbook gently, as if it belongs to a course like ENG 101. It makes a great ride. First three weeks are crucial, and don't let a problem grow bigger and become out of control for any reason.

August 23, 2015

7. FACTS & FACTOIDS - HISTORY & HERESIES

The first chapter in any standard textbook on history of mathematics 'compulsively' deals with mathematics in the prehistoric era. There should be no 'the' before 'prehistoric', as it is always in flux. Anyway, it remains the smallest chapter despite all the paddings and verbiage. Also, the qualifier, 'prehistoric' before 'era' means that the 'reliable' records do not exist or are not recognized in 'pre pre pre' times. In simple words, the history of that subject/topic has not yet taken off.

This mode of thought and approach is what I call a Darwinian model. The origin of everything is assumed to be in the most primitive state of existence. Intelligence is always evolving over eons of time and a myriad of environments. This theoretical construct is very popular in the western academe. For example, it has been applied to justify colonization of major parts of the world since the 16th century.

However, man has been endowed with the faculty of memory - he can recall and retrieve incidents and events from the past. Likewise, there is a collective memory - say, of a family, society, and even of a nation. It is the 'projection' of memory into the future that collectively and individually, certain events and artifacts become worthy of historicity for the next generations.

I have to be critical to this line of thinking too. A homeless person is consumed with his next meal and night shelter. The questions of history and legacy have no meaning for the indigents. It equally applies to all impoverished societies and nations. They continue to live in the dark ages. History is a privilege of the rich and of the victors, who use it to preserve and perpetuate their present.

There is an important offshoot of any prehistoric study, and that is the **historic significance** of a finding - be it of a fact or factoid. Not every ancient piece has historical value. Ironically, any object of mathematical significance is more likely to be lost forever. Reasons?? Mathematically, I take a limiting case at the individual level. It is a fun exercise of recalling one's last 24 hours. One can read a 200-page book in two hours, but a record of the previous day may not even fill a dozen pages. Well, move back to the day before yesterday, and so on. The recalled data drops super exponentially after each day - on the top of all kinds of mix-ups in the happenings.

What is the bottom line? Assume there is an incident of importance for a particular individual. It may lose its shine over a period - like, it happens to one's early research in academe. More importantly, what is of historical significance to one individual or society may not be the same to another individual or society. In view of short public memory and taking the starting point at 2001, the ongoing systematic destruction of the 2000-year old monuments and manuscripts in the plazas, libraries and museums of Afghanistan, Iraq and Syria, provide a perfect example of the relative historic significance in the societies.

Yet, in another limiting case, **the planet Earth does not have enough shelf space** for saving every single historical artifact and monument! If humans lived on forever, then after only 20 years, humans would start pushing each other off the edge of the earth! Man-made disasters - like, certain destructive actions by Al-Qaeda, ISIS and Taliban; and natural disasters - like fires, floods, tsunamis and earthquakes, all perhaps, balance the eternal cycle of creation, preservation, and annihilation - the Divine Trinity!

August 25, 2015

8. CAUTION IN UNDERSTANDING HISTORY

In order to understand the full context of any aspect of life, it is a natural tendency of the human mind to go back to the recent past, and then to the past of the past (mathematically to ad infinitum). I often wonder, if there is such a faculty of curiosity about the past amongst the birds, animals, insects, fish, reptiles and even inside the microbes. It would be wonderful to know, if any other species possesses such an intellectual wanderlust too. To a large extent, the homo sapiens are genetically programmed, besides environment and culture factoring into it.

This line of thinking always strikes a cord in my mind, whenever I discuss early traces of mathematics as given in most textbooks on the history of mathematics. Yesterday, I told my students that one of the fundamental differences between mathematics and history is in the nature of references given at the end of a book and/or chapter. In mathematics books, they are all solid, correct - no disagreement. But in history, the claims may be totally wrong in the sense of even the evidence claimed therein may not even exist.

Mathematicians move with their mathematics globally without disagreeing over proofs of theorems. But historians may have social and political agendas. Consequently, they move in packs, like wolves in the wilderness. From the top, historians, both in academics and freelancers, are supported in certain lines of scholarship or else are ostracized for holding views at invariance with those in power. It is perpetuated downwards accordingly. It comes from the nature of discipline where an event is likely to have different interpretations depending upon the angle it is viewed from.

The number one fallacy in most textbooks on the history of mathematics is the assumption that societies which existed over 2500 years ago, all over the face of the earth, were totally primitive. I stoutly repudiate it. The existence of the monuments like Egyptian and Mayan pyramids provide ample proof of high class engineering existing 4000 -5000 years ago, thus indirectly implying the knowledge of arts, science and mathematics in that particular era and society.

One of the reasons for this mindset is that textbooks on the history of mathematics are written by math professors who are trained to think in hard core mathematics for their doctoral work and professional researches. From a reasonable sample that I have checked, hardly 1 % of math professors have a sense of history and political science. Ironically, math faculty in the top tier universities are far more one-dimensional - that the complement of their math expertise is nearly empty. The examination of the intellectual breadth of math faculty in various university tiers is a research topic of its own kind. At least, I have not encountered a math professor who has an equally solid background in history too.

In the pursuit of a deeper and wider understanding, generalization of a conclusion is a common intellectual resort. Mathematicians are meticulous and careful about it. Anthropologists and archeologists tend to go to the other extreme, as they can sweep over miles of territory from, say, 10,000 sq ft of excavated piece of land.

Yesterday, I exhorted my students to be a part of the following mental construct. Say, after 5000 years from now, people want to understand the inhabitants of Las Vegas Valley. If a bunch of archeologists dig out in the 'west side' of Las Vegas, specifically bounded by Las Vegas Blvd on the east, US 95 from the west, Stewart Street from the south, and Lake Mead Blvd from the north, then they would conclude that the Las Vegas Valley was inhabited by the most primitive people. However, if another group of archeologists dug up any piece of Las Vegas Blvd, say, between Mandalay Bay and Stratosphere, then they would project a picture of the most scientifically developed society!

In conclusion, I cautioned the students to be mindful of any conclusion drawn in a history textbook on any subject. I stressed that the non-literate societies, believers in the magic of numbers, and communities practicing divine powers, as mentioned in the first chapter of the textbook, have always existed in every nation. They can be seen today in the pockets of urban affluence, in the jungles, islands, and remote regions of our vast earth. The bottom line is that they do not constitute the entire spectrum of human life.

Sep. 01, 2015

9. VIGILANTES IN HISTORY OF MATHEMATICS

Have you ever wondered about having your name on an object, especially if it ends up being placed in a famous library, institution or museum? By this, you achieve a piece of immortality by association only - for the heck of finding a mud-covered nugget in a garage sale, rummaging in a junkyard, or just turning over items in boxes stowed away in the attics of grandparents. Americans are the biggest collectors of goods. Here are the historical items, which are perhaps mined from the trash and junk of everyday life.

1. **Dresden Codex of the Maya**. Dresden is the name of a city in Germany. Codex means a code or codified object. It is worth Googling for details. My point is that it dates back to the 13th century. Somehow, it has survived the 16th century of total holocausts of Mayan civilization perpetrated by the conquering Spaniards. The destruction included the native populace, their temples, literature and scriptures. The codex surfaced up (??) after 200 years and may have been recognized for its rare appearance by a curious local or Spaniard (??). It must have passed through a few hands before the library in Dresden purchased it for its museum.

2. **Rhind or Ahmes Papyrus** is a scroll/roll purchased in Luxor, Egypt by Henry Rhind, a Scottish antiquarian (I would call him a curious traveler like myself!) in 1858, when the European colonization of the world was at its peak. Ahmes is the name of the scribe who copied it in 1650 BC, but it is not known who the author was. Anyways, it remained dormant for at least 3500 years! How?? Try to get a grip on this time span.

3. **Moscow or Golenishchev Papyrus** was purchased in 1893. No one knows where it remained hidden for centuries, and who first saw a value in it. It is housed in the Pushkin Museum of Fine Arts in Moscow. Don't you wish that you were there too, and strike gold? The national museums of London and Paris are not behind in claiming similar rolls. Note, semi-mathematicians and scholars like Champollion come into the picture at the very 'end' of the game. These scholars remind me of vultures and jackals - called scavengers, who approach the remains of a carcass after the lion has had his fill of the kill to his satisfaction!

In the above context, two things come to mind. Today, the modes of travel are exponentially far more in availability and faster as compared with what existed, say, 200 years ago. However, mobility doesn't make one observant! In life, one can be mentally mobile by introspection and contemplation while remaining physically stationary, and yet bring out the nuggets of life. Theoretical researchers do it all the time. If you are traveling fast, then you do miss life around you. Being observant while in motion physically is very challenging.

For example, in a supermarket, I literally don't see the items around me and miss throw-away deals. The reason being that I enter a store with a list of items on which I am focused. Unless an item next to the one on my list catches my eyes, I would normally miss it. On the other hand, my wife is so good that, for example, she would right away spot any new item in the houses of her brothers, whenever we happen to visit them.

For students of the history of mathematics, the faculties of alertness of both mind and body are to be trained. If you browse the history of mathematics textbooks in the library, you would be shocked to realize that the ancient history of mathematics hinges on only these 4-5 objects. How fair is it to stop it there? It is ridiculous to believe that that was the extent of mathematics of antiquity. Another point is that we need only to look at other sources of mathematical information, or search for such items, which may get scarcer with time.

Finally, if mathematicians are not going to hunt for the mathematical 'treasures' of antiquity or medieval times, then no one will do it for them. It is time for the mathematicians to team up with archeologists, anthropologists, historians, and even sociologists in putting a composite mathematical picture of antiquity together, especially if (or when) an ancient artifact of 'mathematical' significance is discovered.

Above all, keep your eyes (including the **third one**) always open when you are in a new place!

Sep. 22, 2015

10. EXTREME STATES OF MATHEMATICS

"Give an example of a society or nation, where mathematics has flourished in isolation from other disciplines." This is the very first problem that I phrased and included in the **Test # 1** in a course on ***History of Mathematics*** (MAT 714) given two days ago. I did anticipate all kinds of answers. But, after grading this problem, I felt that all the nine students had missed the full intent of this question. It does happen in a 'testy' environment, particularly in an open course like this.

All the nine students are graduate students - including two in the doctoral program, who are taking this course as an elective. Five students are doing MS in the ***Teaching Concentration,*** where this course is required. One is in the MAS (Master of Arts in Science) program, housed in the office of the Dean of College of Sciences (not in any department). The last one is a math PhD professor in the College of Southern Nevada. At age 60-, she is taking this course simply to expand her intellectual horizons. Such mathematically aware students stretch my intellectual limits too.

In regular lower division math courses, I give out solutions of the test problems when the tests are returned, but allow some flexibility at the upper division level. The graduate students are supposed to figure out the missed solutions afterwards, as this is a part of their training in problem solving and independent thinking. Anyway, I did ask myself - what would be my answer to this question? That is the genesis of this reflection.

A good part of the answer lies in one's perception of what is mathematics? It is too early for it to crystallize in the minds of the students, irrespective of whether or not a student is a genius. Appreciating mathematics is a part of a maturing of the intellect. It even transcends one's ability to solve difficult math problems or prove good theorems. After living off mathematics for nearly 55 years and lately making mathematics a paradigm of my life, I have concluded that mathematics is a language, a tool, a conceptual model, and a type of thinking that is completely deductive. Let me examine each category briefly, when mathematics flourishes.

Since the 16th century, mathematics has been recognized as the language of the sciences. I have not forgotten my awe at the power of math that I noticed during my college days (1955-59), while studying chemical formulas of compounds and chemical equations - both in inorganic and organic chemistry, and mechanics in physics. To make it short, when mathematics flourishes, the natural sciences bloom - and vice versa.

Looking at mathematics as a tool in engineering, the skyscrapers, ocean liners and supercomputers, mile-long bridges, weapons, space and earth exploration are strongly related with math in action. One can see it both directly and conversely. Any society or nation in the past or present that has built structures like the pyramids or Freedom Towers was/is advanced in engineering/architecture and math. Therefore, the engineering, financial and medical disciplines are never isolated from applied or computational forms of mathematics.

In the US and countries of Western Europe, there is witnessing a fusion of mathematics and arts at the macro levels. Mathematical concepts are being blended in arts - whether in painting, dance, music and sculpture.

What are the symbols and symptoms of mathematics as deductive thinking? Its answer lies in a deeper grasp of mathematics. Deductive means closely connected, which implies being organized on a large scale. Mathematics may be pursued by a genius - like, Ramanujan, in apparent isolation, only for a while; but it only flourishes when it is brought into a collaborative framework. This large-scale organizational nature of math is reflected in everything super-big - whether in the universities, corporations, sport stadiums, and banks! Looking at it contrapositively, one can see this point in many nations that are labeled as underdeveloped!

When mathematics pervades in a society, even philosophy, a discipline far removed from mathematics, becomes analytical/mathematical, as evidenced in the western philosophy going back to the Greeks of the BC era. They looked at mathematics and philosophy as inseparable entities of scholarship. This is opposed to the eastern philosophies, where absence of mathematics is covered with its amalgamation with the elements of divinity (or God) in it. Philosophy then becomes esoteric.

Let me add that mathematicians do not have a monopoly on numbers, fractions and fundamental geometrical shapes. They are in the human genome. However, one may raise a question: are there any human endeavors, which do flourish while being miles apart from mathematics? Yes, certain forms of 'free' poetry, arts, music. dance, and medicine. They are erroneously called tribal and primitive, as only their simplistic forms are placed before the public.

A type of creative work comes out under the influence of some mind-altering drugs and weeds, mushrooms, or opium found in nature, John Keats (1795-1821) composed his time -immemorial poetry under the influence of opium. The kaleidoscope artworks are produced similarly. It basically says that creation of mathematics and such arts stem from diverse states of brain cells. A bottom line is that there are specific **medications** (from without) and **meditations** (from within) which can equally alter a human mind at its molecular level.

There are individuals known throughout history including Jesus, Buddha, Confucius, who have taken the domain of spirituality to heights far beyond any mathematical construct. Here comes an intellectual cusp or bifurcation point. **Anything, once mathematized, becomes communicable to the next generation,** In more concrete words, the spiritual achievements of the ***sufis***, sophists, ***rishis***, ***yogis***, Zen masters cannot be communicated in a format whereby one could systematically study them online at one's leisure and become, say another Jesus! On the other hand, people do study math and sciences, and can go beyond the giants in the field.

Mathematics is like an inverted shiny granite pyramid with a broad base and constantly growing vertically up. Spirituality is a union of similarly inverted pyramids each sitting at individual points - defined by the 'high souled' individuals. However, there is a latent connection between any two such persons - say one, spiritualist and the other, a mathematician or scientist. My conjecture, based on some 'direct' experience, is that in one case, mathematics is 'internalized', but in the other mathematics remains external! In one, it is latent, but transparent in the other.

Finally, in our constant desire to compare or rank any things in life in terms of higher or lower, I would not venture to do it!! Both 'paths' give me moments of feeling 'high' in life. I am fortunate for having tasted them, glimpsed them, heard them, smelled them, touched them, and have absorbed them in my being. This realization in itself is a pathless path!

Coming back to the Test Question #1, is there a one-line answer? Yes, there is one - in fact, in one word, that is. NO! Also, I do not have an example of such a society or nation. The set is empty so far! Furthermore, a proof by 'contradiction' is interesting to play around with!

October 23, 2015

COMMENTS

Dear Prof. Bhatnagar, I tried to formulate my opinion somehow, but it seems that I failed. The question is too challenging, and the answer does not exist. For those who want to speculate on this topic, it requires knowledge of history (not only the history of math!) together with the ability to think deeply. That is too much for many people. So, I could easily fail your test. At the same time I can imagine a person who will spend not all the time on a test, but many years of his or her life if he or she will be caught by a question like this. It depends on the person.

It came to my mind that it could be interesting to discuss the connections between the so-called "pure" and applied math. How people came up with ideas that were ahead of their time. Do such cases (examples) have any reasonable explanations? How does it happen that some achievements of pure math may become applied on some day? But I am pretty sure that in your course you consider such topics.

I wouldn't probably totally agree that math is a form of "deductive thinking". Sometimes, intuitive thinking can be crucial. Again, I am not sure that my "feedback" may be of any help. But thank you for sharing your thoughts. Have a nice weekend! Sincerely, **Viktoria**

11. UNIQUE TWINE OF MATH & HISTORY

Even after five decades of full-time teaching, I still get butterflies in my stomach while walking to the first class of the semester. It is due to anticipation and expectation – both on my part (real) as well as on the part of the students (imagined). I have no idea of students' expectations of me and vice versa too, as I do not know them yet. On the first day, there is a list of things to be explained about the 'business' side of the course – mandatory. Yet, in order to set the tone of my teaching style, I do dip into the course material. I like to 'marinate' the students and tune them up, like one does to musical instruments before playing them. The idea is to align them with my philosophy of teaching as soon as possible.

Having taught in five overseas assignments, I find that there is nothing like the first day on the campus of an American university. A genuine atmosphere of merriment and curiosity is seen on the faces of people buzzing around. From my office in one of the cluster of 20-year old modular (temporary) buildings, I walked out between the Beam Hall and the 5-storey (under-construction) new home of the College of Hotel Administration, UNLV's perennial academic showpiece. My thoughts went off to the year-1983 when the Beam Hall was built. From a historical perspective, it is safe to say that 90% of the university structures have been built or extensively remodeled since my joining UNLV in 1974. Once the Hotel College moves into its new home, there is a possibility that the Mathematics Department may shift to the Beam Hall –its fifth move since 1957! UNLV is one of the few major universities in the US, where the Mathematics Department does not have its own building – nor, a permanent home.

My first class of this semester is **History of Mathematics** (Math 314), which is scheduled in the Carlson Education building on the north end of the campus. After crossing the busy east-west pathway between the Humanities building and Lied Library, one meanders through the Don Baepler Xeriscape Garden. Don was UNLV President when I joined! Later on, he served as the Chancellor of the NSHE (Nevada System of Higher Education) and ran the Harry Reid Research Center for 20 years. He died a few years ago.

Coming out of the garden, I swung right. On my left was a rotunda, which was the first home of the UNLV library until 1980. It then looked like one of the Las Vegas casinos, as constructed through the 1970s! On the right was the John Wright Hall recently rebuilt after total demolition. However, the name is retained, as Tom, John's son, is an Emeritus Distinguished Professor of History too. Dina Titus, Tom's wife, an Emerita Political Science Professor, has been in state and national politics for three decades. The main north-south academic mall is awesome with green lawns and tall trees - majestically aging. My kids rode their bikes here, when we lived, just across UNLV, in the University Park Apartments (half of them are now demolished) from 1974-77. This academic mall is an oasis in the desert.

On the left side is a rectangular building, which underwent extensive remodeling with financial support from James Rogers, a philanthropist and NSHE chancellor. From 1982 until 2000, it was the second home of the UNLV library. The space on the west side, between this building and the educational one, opens out onto a striking horticultural vista. With such thoughts wafting through my mind, I entered into the Carlson Education Building, which was completed in 1972. Its architecture has protective features against any nuclear fallout – a reminder of the Cold War/Star Wars era. The Mathematics Department was housed in this building until 1992.

In my 20-minute extemporaneous monologue, I told my students that I could see the history of anything. At the same time, mathematics is a window and a paradigm of my life. I understand and appreciate things, which are expressed in mathematical forms. It is often said that God, if there is one, must be the greatest mathematician. However, I go further than this - that God has to have a mathematical form!

I explained how a professor of history could not teach a course on History of Mathematics, as he/she hardly knows calculus. On the other hand, a mathematician's mind is inhibited from having a sense of history. Mathematics is objective and binary, whereas, history is subjective. Proofs in mathematics are like carvings in stone. Proofs or facts in history are always in flux. In order to set the tone of disciplinary distinction, I assigned the following problems on the very first day:

#1. "Find out the name of your grandparent's grandparent (great great grandparent - one of the sixteen). Not everyone may have a biological child, but everyone has two biological parents, four biological grandparents, and so on. Provide some 'proof' of your claim. You will gradually understand that 'proof' in history is not the same as understood in mathematics."

#2 "Use **any crude mathematical model** to estimate the number of people who have lived and died on this planet Earth during the last 2500 years. Name a few persons who lived **at least 2500** years ago. Turn in both of your findings on Monday, 01/23." These exercises will develop students' perspectives on history in general. It would be helpful whenever historical scenarios are encountered.

I described my journey into the diverse areas of mathematics and history. During my schooling in India, I became so proficient in arithmetic that I resisted learning algebra. Eventually, I appreciated the power of algebra over arithmetic, but then I resisted learning Euclidean Geometry and my frustration over proofs of propositions. All these moments are etched in my mind. Now, I claim that in the entire gamut of mathematics, there is no body of mathematics that is so beautiful, complete and useful, as Euclidean Geometry is.

In college, analytic geometry made a new believer out of me. I recalled my excitement when at the age of seventeen, I saw the sheer power of mathematics in the study of Inorganic Chemistry and Newtonian Physics. I studied mathematical physics through my BA and MA in India. Mathematics has been changing me, as I continue to explore mathematics. Also, in a nutshell, my mathematics journey has rising peaks and valleys – always lifting my soul.

I told my students that I hated history in high school for its emphasis on memorization. I cannot remember even two lines of a song, though I am humming old tunes most of the time! In India, due to curricular reasons, history was miles away from mathematics and sciences. My love for literature and philosophy brought me close to the makers and shapers of world events, and I delved into history and political science. One cannot go too far in mathematics by one's efforts alone; however, social sciences are different.

By the age of 40, mathematics and history seemed to have found a fusion in my consciousness. In the 1980s, I taught two short experimental courses on the history of mathematics. My approach towards the history of mathematics is to view it as a subset of history in general. I may be the only professor of mathematics who has written a book on Indo-US historical issues. Its name, ***Vectors in History*** (2012) is mathematical. ***Darts on History of Mathematics*** (2014) is a mathematical corollary of the ***Vectors in History***.

In the context of my books being written in English, I told the students that my mother tongue was Hindi. The second language that I learnt in elementary school was Urdu, a derivative of Persian and Arabic as far as the script was concerned, but it sounds Hindi when spoken. The third language that I learnt was Punjabi, the regional language of Punjab, where I grew up. English was my fourth language, which was started in the fifth grade – with no emphasis on oral communication with others. It was plain reading and writing with a touch of grammar.

This is the gist of my today's 20-minute monologue. Monday was a holiday in remembrance of Martin Luther King Jr. Since classes generally meet on two days a week, and mine are on Mondays and Wednesdays, the first week of instruction has already become a part of history!

January 18, 2017

COMMENTS

Satish, When you are doing history of our math department, you might do a comparison between the number of buildings added to the UNR campus and the number of buildings on the UNLV campus starting in 1968. Then ask who paid for the buildings. My suspicion is that we (Clark County) paid for the buildings. Be careful to note when federal funds are involved. The suspicion is that there is gross inequity best illustrated for us by the fact that UNLV started its PhD in math before UNR did theirs. Yet UNR has a math/ science building and as you point out, math at UNLV has no "home" and is in temporary, inadequate quarters. Unlike your personal math history, mine began in the third grade when I invented fractions in order to answer the teacher's question: what is half of 15? Keep on writing. **Paul Aizley**

12. IMPORTANCE OF CLASS PROJECTS

A history of mathematics (HoM) course is a unique course in any college catalog of mathematics courses. This applies to every facet of the course. Its instruction is challenging in the sense that a typical math instructor is always facing the students, not talking to the writing board. Unless the instructor has some sense of global politics and general history, HoM would remain a social disconnect. The choice of course material is no less challenging for the students to have a holistic view of HoM. Available materials present a narrow view of HoM, and they especially fall short on HoM pertaining to the African, Asian, and South American countries and cultures.

The teaching/taking a HoM course without any hands-on experience of individual history projects is like teaching/taking a hardcore science course without providing/experiencing flavors of its lab. In the sciences, the lab associated with theory helps in the development of a scientific bent of mind. In HoM, individual projects develop students' sense of history by gathering facts from reliable sources (no all-Googling!) adn interpreting them

The students choose their topics by the end of the seventh week, and have five weeks to work on them. They make five-minute presentations before the final submission in the last week of the semester. The project covers 20% of the grade and is limited to 1000 words. In the previous offerings of HoM courses, the students have tackled the lives of mathematicians, histories of the courses - from remedial to lower division courses; profiles of graduate students, gender based data on majors and minors, faculty and administrators etc. The only limitation is that the student must like his/her choice of the project and take full ownership.

Historically speaking, the idea of the HoM projects synchronized with the 50th anniversary celebration of UNLV in 2007, when I was teaching a HoM course. Four years ago, these projects were classified, compiled and organized in two 3-ring folders. The Special Collection Division of UNLV's Lied Library (digital) readily accepted the entire material. It is now available to any curious person and institutional researcher who is curious about the Mathematics Department in its 'teen' years.

March 03, 2017

13. PEDAGOGY OF HISTORY OF MATHEMATICS

History of Mathematics (HoM) can be the most enriching course for the students provided the instructor is intellectually well rounded - mathematically, historically and sociologically. Teaching a HoM course always energizes me with new ideas, which are subsequently anchored into several reflections. However, the writing of this reflection sprang up today while grading the biweekly reports of the students. Pedagogically speaking, this grading aspect of the course has brought something fresh for me; whereas, in the class, I steer discussions.

Any course in HoM is far from being a hardcore mathematics course - like math done in a Pre-Calculus, Calculus, or Discrete Mathematics course. The prerequisites of an HoM course are purposely kept nebulous. For instance, UNLV Math Department offers two HoM courses - the one at the undergraduate level (MATH 314) has Calculus II, as its prerequisite, and the other at the graduate level (MAT 714) has 'graduate standing' as its prerequisite.

At the same time, an HoM course is not a course in history per se. The history courses offered in any History department focus either on a period - from contemporary to an interval, to back in time a few centuries, or on a region, anything from a city to a nation. The focus could also be on the history of a single event - like, the Holocaust or Marxism. My main approach to an HoM course has been to examine HoM as a subset of history in the most general format. It comes naturally to me as my interest in history, literature, religion and philosophy has been on tracks aparallel with mathematics since my college days.

In a mathematically homogeneous group of students, an HoM course may be 'taught' as the history of a bunch of mathematical topics and problems. In an undergraduate course, say, something like the history of the quadratic formula or trigonometric identities may bog the students down in muddy techniques and special cases with little relevance to the present. Likewise, I can visualize the exotic sounding history of the **Prime Number Theorem** and the **Fundamental Theorems of Calculus and Algebra** getting mired in technical details. For instance, the first proof of the **Fundamental Theorem of Algebra** was essentially Gauss' (1777 - 1855) doctoral dissertation - running into some 20 pages.

One has to bypass or waive the mathematical rigor in an HoM course. The only thing one can do is to mention the names of mathematicians and their research papers preceding the proof of these theorems. There is a textbook on great theorems in mathematics, but I am not sure how it would cater to an undergraduate HoM course. As far as the choice of mathematical topics is concerned, it is wide open. What is the good of only knowing the names of mathematicians alone unless their lives are not examined in a broader context of their institutions, societies and political systems etc?

In conclusion, there is a 'similarity' between history and mathematics. Everything has history connected to it - a story of its past, be it short or long. There is mathematics behind everything too. Anything can be examined from a mathematical angle, or be seen as having a mathematical model. Nonetheless, any aspect of HoM becomes interesting with the presence of the human element in it. Therefore, my approach to HoM is to bring in the human drama as much as possible. In a nutshell, an HoM course becomes livelier once history is shaved off its dates and mathematics of its epsilons (ε) and deltas (δ).

March 23, 2017

14. MINDS NEED FRESH AIR TOO!

The US lifestyle has been evolving as I started witnessing it fifty years ago, when I came from India to the US. Bringing fresh air inside the homes and offices is a figure of speech. Fresh ideas are like fresh air too. New pedagogical injections are necessary in staid classroom instruction and even in individual research. Any change in scenario and perspective provides a stimulant to the entire process, where the minds are engaged in any proportion.

This line of thought was triggered yesterday while I was grading a test question, on the merits of the biweekly reports, set in the final exam of History of Mathematics (HoM) course (Math 314). This course is required only in the Secondary Math Education degree program. However, in a class of 27 students, 21 of them are from other disciplines - physics, math, engineering, computer science, and economics.

Since additional reading and library/online research are encouraged, I requested a reference librarian (having a PhD) to give a talk for 20-30 minutes on what was currently out there on the internet in the name of information and scholarship. Googling for any kind of inquiry is becoming a reflex action. This lecture was fixed on the very second day of the semester. With examples, he informed the students about the reliable sites, but also cautioned them about fake information posted on some sites.

The second lecture was scheduled in the eighth week of the semester. A HoM course is neither a typical mathematics course, nor is it a regular history course. One has to know at least calculus. At the same time, one should be open to different points of view, which is not a characteristic of a normal math student or faculty. The world of mathematics is largely black and white. Incidentally, the Chair of UNLV's History of Department came over to face my students. Yes, quite a few history faculty were reluctant to address the questions concerning scholarly methodologies and facts in history. The crux of his remarks, "There is a history of everything" hit like a home run/sixer with the students.

The third speaker was an emeritus professor of computer science, but with a Berkeley PhD in Differential Geometry. We have known each other for 30+ years since Computer Science was a part of the Mathematical Sciences Department until 1982. It took me a lot of cajoling to convince him on the merits of giving such a lecture. I said, "At the age of 75+, such talks are our way of giving back to society what we have accumulated - be it money, objects, varied experiences, and knowledge." Generally, mathematicians are not good at public speaking. Finally, I said, "You just tell the students the story of your life". That is what exactly he did and in his natural demeanor. The students were swept away by his candidness. This talk was given two weeks before the instruction was to be over for the semester.

Guest lectures have little room in regular math courses, particularly at the undergraduate level. Also, any guest presentation must supplement and complement the class material - directly and indirectly. Personally, I can still recall the nuggets gained from such guest lectures heard in India decades ago.

What do I do for the guests in return? Of course, I offer to return the favor in their classes. However, I have yet to speak in a history or political science class. It is due to the image of social disconnectedness that mathematicians and their discipline project. So, I have individually treated them to an Indian lunch buffet in a restaurant near UNLV campus. We eat and talk over the finer points of our disciplines. Intellectual enrichment has no boundaries.

May 15, 2017

15. BRICKS AND BOUQUETS ON *DARTS*

For me, the writing of a book and enjoying it is like fathering a child and loving it. Most parents (modulo a set of measure zero) do love their children, but very few people write books, and much less are proud of them. The ***Darts on History of Mathematics*** is an outgrowth of my miscellaneous articles and reflections on the history of mathematics (HoM) that I have been writing since the 1980s. Most of these reflections were prompted by my teaching of the undergraduate and graduate courses on HoM during the last 15-20 years, and attending the HoM conferences and sessions in the US and abroad.

After the publication of the ***Darts*** (for abbreviation) in 2014, I casually used it as a supplementary book in a HoM course. Because of its no-sequential style in terms of topics, eras, cultures, and regions, it is not meant to be used as a textbook. On the other hand, there is nothing like the ***Darts*** as far as its use as a supplementary book is concerned. That has been my opinion, which may have an author's bias. Therefore, seeking validation from the students was natural.

On May 03, 2017, the last day of the spring semester, I asked my 26 (out of 27) students of HoM course (Math 314) to use one side of the sheet for the evaluation of the textbook. ***Math Through Ages*** by Berlinghoff and Gouvea (2015) and the other side for the evaluation of the ***Darts***. On the sheet, the spaces were marked out for the writing on just two ***Strengths***, and only one glaring **Weakness** of the two books. I stayed out of the classroom until everyone was finished. The names on the sheets were optional.

Here is the summation on the strengths and weaknesses of the ***Darts*** from the students' perspective. The numbers in the parentheses are the numbers of students who thought alike on a particular point.

STRENGTHS

(2) 1. Exposes to a variety of cultures - uncommon in most textbooks.
(2) 2. Stories of students are easily related by the students.
(2) 3. Unique perspective on HoM by bringing other elements of life.

(2) 4. Elements on HoM like the study and research in HoM in India.
(3) 5. Enjoyed reading about a professor passionate about HoM.
(2) 6. The idea of the book is very interesting.
(3) 7. The format and topic of each reflection are engaging, informative and often thought provoking.

8. Each Reflection is different. I don't have to read it from the beginning to understand what is going on.

(6) 9. Tons of first-hand experience, personal anecdotes that relate to the topic at hand.

10. Material is relevant to any college student with plenty of advice.

(2) 11. Short and to the point.
(3) 12. Feedback of readers is included in the book - provides unique insight.

13. Sociological, political and linguistic impact of math - interesting thoughts.

(4) 14. Contains information that many do not think about.

15. Succinct - does not drag on.

16. ....profound pieces of information.

17. Gives a lot of insight on things you would not find in a normal history book.

(2) 18. Seeing other people's thoughts other than the instructor.

19. Reading your book with your personal insights gave us more knowledge of who you were.

20. Showed a more human/personal side of history.

21. The reflections made me reflect and think about history and the present.

22. The use of flowery language was unique.

23. Since many ideas revolve around UNLV, it feels relatable/applicable to me as a student.

24. He touched unfamiliar topics and exposed students to new ideas.

25. I love reading quick thoughts on a topic that are triggered by some seemingly random incident.
Each Reflection is just thought up in the moment and I like seeing how things are connected.

(2) 26. Getting to see into your mind and seeing how you make connections was inspiring to me.

27. Most books are biased and informative and it was really nice to have a book that discussed thoughts and ideas.

28. Talks about Math through experience.

29. Complements the class very well.

WEAKNESSES

(5) 1. Repetitive
(2) 2. Spelling and grammatical errors
(2) 3. Could use a re-ordering so relevant reflections are in groups.
4. Parts of the section are tangential to the main topic of the section.
(3) 5. Each Reflection should be numbered for quick navigation and search.
6. Slightly seem all over the place.
(2) 7. No unifying theme - many points, but few strings to tie them all together.
(3) 8. Very unsure what I was reading about at times.
9. More information on the topics needed.
10. Books jump around a bit on topics. It needs a clear goal or better organization.
11. Some chapters could have been left out.
12. Talks more about history than math.

MY BRIEF RESPONSE

Some of the positive comments are even beyond my expectations. Yes, at the age of 70, when I started writing non-fiction books, one of the objectives has been to add newness and freshness in respective fields. I intend on beating my own benchmarks.

Since the writing of the very first book, I have taken full ownership of my book. No ghostwriters, consultants, proofers, editors, or designers. The main reason being that I have fallen in love with the English language again. The only way to master English is to do the proofing and editing of my books – it is a journey into my soul. I am getting better at the language and so are my books! Naturally, some stubborn errors from the typos to spelling would go undetected. You can only catch so much of your errors with your own eyes!

My writing style is my signature. It may not be pretty for some, but that is the way life is. My ultimate focus is on my ideas, which are marinated and brewed for a while. Thus, I am open for a contest of ideas. Furthermore, when a book can be read from any page, then it has to be independent in all respects - thus repetitions are unavoidable. Material has to appear disjoint. In a tweet age, such a format of material is ideal.

May 18, 2017

16. ORAL HISTORY AND MATHEMATICS

Oral history occupies a respectable pedestal in the US society. One of its offshoots can be traced to the Afro-Americans gaining political clout, and their resolve to have their historical roots well defined. Naturally, this has to go back to the 17th century, when young men and women were lured and captured from the regions of western Africa, and then shipped in chains and shackles to the US shores. These slaves lived under the harshest living conditions. The only recourse to their legacy was the passing on of their stories from word of mouth, as they were barred from attending any school. Thus, oral history remains integral to the lives of the Afro-Americans. In 1976, President Gerald Ford declared February as the Black History or Afro-American Month, though its checkered history can be traced back to 1926.

For a cultural context, India has a long history of both oral and written practices. For instance, the Hindu scriptures were both memorized and written down for millennia. Also, the wandering bards sang social ballads and a small group of artists enacting skits on street corners. In every town and village, there was/is a ***chopal*** (central place), where the elders would gather and leisurely discuss social issues. At the other extreme, such gatherings would turn into ***gupp-shup*** (means bull shitting). Therefore, facts in the oral history could be individual opinions - true, false or semi true/false, as they do not undergo any scholarly cross-checking. Nevertheless, it is better to have something rather than nothing.

Oral history is getting a new push in the digital age, which has seen families shrunk, divided and dispersed - thus, creating a new i-alienation and break-ups in the society. It can be stated mathematically that one's radius of social contacts is inversely proportional to the speed of i-communication. For instance, only 70 years ago in the US, one could see three generations of a family living together under one roof. But, the present generation of the millennials has no time even to get married - much less have children. I often urge my septuagenarian friends to start writing the stories of their lives, which may serve as a kind of resource material for their grandchildren. But where are the grandchildren in the first place?

The writing of this reflection is triggered by Claytee White, who happens to be an Afro-American, and is the Director of the Oral History Project housed in the UNLV Libraries. On September 06, Claytee will be a guest speaker in my course on ***History of Mathematics*** (MAT 714) that I am teaching this fall semester. I have known her for the last three years, and as a part of this project, she has interviewed me about Indians in Las Vegas. It is worth adding that Panjab University, Chandigarh, my alma mater (MA, 1961) in India, has an Oral History Project that is housed in the office of the Dean of Research.

There is an undeniable understanding in the study of the rise and fall of human conditions over the centuries - all due to written records, as opposed to oral traditions. Thanks for the written history of every discipline, at least, the wheels are never re-invented. On the questions of mathematical nuts and bolts of oral history, here are the following two questions, which I keep as the focal points:

1. **What is the role of oral history in the development of hardcore mathematics?** This question is easily disposable, as no society of the past or present has made any impact on mathematics, with oral communication being their only mode of communication. From its practitioners, mathematics demands preciseness and understanding, never less than 100 %. On the other hand, an interesting fact in a theory of oral communication is that at least 20 % of the content is lost when a person transmits it orally to the other person!

2. **What is the role of oral history in the History of Mathematics (HoM)?** The domain of HoM, being a subset of history in general, is much wider in terms of its relevance. It includes people, places and institutions associated with mathematics. As a matter of fact, the hardcore mathematics remains in the background like a theme song. Briefly speaking, through oral communication, one can collect mathematical stories on researchers, teachers, curricula, textbooks, instruction and evaluation criteria etc. In HoM, my area of research is to explore the necessary and sufficient conditions (mathematical jargon) for the development of mathematics - be it of an individual, society, institution, or nation at large.

Interestingly enough, the first assignment that I give to my students in an HoM course pertains strictly to oral history. Consequently, we also explore what kind of HoM one would talk of in an ancient society that goes back to at least 5000 years, and its merits. Thus, the study and impact of oral history is all up in the air.

August 26, 2017

COMMENTS

Dear Professor Bhatnagar, Your latest reflection on a possible oral history of mathematics is very interesting. As you seem to suggest, an oral history of a family is much easier to do by an elder in a family and it would be welcomed by most younger members of the family. This has been my personal experience. I am the eldest grandson of my grandfather and for many years, I am also the eldest among all the living members of our family.

I retired at the age of 60 in 1995 but immediately got involved as the main organiser of the International Math Olympiad in Mumbai in July 1996. At that time internet connections were available in India only in the four metros. As I lived in Mumbai from March to August organizing IMO, I learnt how to send and receive emails. Without an e-mail facility, we could never have been able to organise IMO. That work was over in August and I returned to Ahmedabad.

Fortunately, the Internet came to my city in October 1996. I immediately went to Bombay, got a PC for myself and joined the Internet community. Fortunately, all my siblings and cousins were in Metro cities or in Western Countries, so I began writing to them episodes from the history of our family. My messages were very well received by all and I continued to excavate various episodes from my memory so the series goes on. One of my daughters-in-law recently told me that my whole family (my children, cousins and their children) are discussing how they can honour me for the service I have rendered!). I believe the very attempt to honour me for this service is the biggest honor ever received by me. Sorry for the message that is longer than I intended. **ARUN VAIDYA**

Satish, nice article on the oral tradition. Claytee does a great job. Sari and I have been interviewed for both Jewish History of Nevada and the ACLU-NV history.

QUESTION:

How does the department like the current facility - The Central Desert Complex? Does anyone actively say the department needs new facilities?

Would any members want a School of Mathematics with at least three departments - Pure Math, Applied Math, and Statistics? Probably a fourth department for medical mathematics or some other connection with biology? **Paul Aizley**

I'd be especially interested in oral history projects that attempted to get important mathematicians to talk about their experiences and ideas. Given the current fashion for extremely terse writing, we lose a lot of "informal" information and ideas. **Fernando**

17. OPTIMIZING HISTORY OF MATHEMATICS!

About US politics, it is commonly said that all politics is local. I have lived in the US since 1968, but I never understood the depth of its politics until recently. Gradually, it began to dawn on me as to how in the US, political culture sprouts from within family environments. This line of thinking was triggered last week by a corresponding question about the study of History of Mathematics (HoM), a course that I am currently teaching this semester. I prefer to examine HoM as a subset of history in general. Therefore, when approaching the study of history in an analogous manner, the history of one's family becomes important, nevertheless. Any mathematical nuggets coming out of it is most welcome.

Granting the point that history is as local as politics, then a subsequent question is about the chronological study of any contextual history. There are various aspects of present and contemporary histories. Here is a popular saying: ***those who cannot remember the past (history) are condemned to repeat it***. By the way, this quote is far more popular than the name of George Santayana (1863-1952), who distilled it intellectually. Incidentally, Santayana and I were born on December 16, and we share diverse intellectual interests too!

Every textbook on HoM begins with a chapter on its **Prehistory**. (Uta Merzbach's HoM textbook calls it **Traces**). Prehistory seems to suggest as if history has a beginning - this is far from being the case! Human history has existed for millions of years, wherever and whenever people walked on this planet. The crux of the question is that from a collective point of view, tangible records of any aspect of social life do not survive after a couple of millennia. That is the reason; such an era is called 'Prehistory'. Understand, mighty Time is the Ultimate Destroyer of all records.

In order to bring this sense of history into focus, in the first two weeks into the semester, I ask my students to estimate the number of people who have lived and died on planet Earth during the last 5000 years. Every estimate is ok, provided there is some explanation for it. It is a nice exercise in crude math modeling too. The two follow-up questions are finding the names of the people who lived between 3000 BC to 500 BC, and between 500 BC to 33 AD. The year markers are simply chosen for their roundness, and familiarity with the Gregorian calendar.

The investigation of these questions brings a reality check on history too. If one goes too far back in time, then in the limiting case, there is nothing concrete left to lay one's hands upon. At the other end, there is so much that happened yesterday that any event has to pass some litmus test of significance in order to qualify as a piece of history. The ideal approach is to extrapolate Santayana's quotation. There is a collective dementia in every society, and it begins to set in after 100-150 years. This time interval is for the purpose of generating a discussion - nothing sacred about it.

September 06, 2017

18. PROBLEMS IN MATH PLUS AND ER!

It has been a month since I circulated the following problem to math graduate students, assistants (doing master's or PhD), and part-time instructors (PTIs have a master's or PhD) - a total of 50 persons:

Show that $27(\square^{4} + \square^{4} + \square^{4}) \geq (a + b + c)\square^{4}$.

In order to promote some action in this exercise, the first person to submit a correct solution was to have lunch on and with me at the **IN 'N OUT** burger joint, which is 400 yards away from my office. Here is a little more than a report card on this mathematical excursion.

No one submitted a correct solution - in fact, not even a wrong solution! One of the PTIs, known to me for nearly 40 years, plotted two graphs for a large number of values of a, b, c - knowing that verification is not a proof in mathematics. Besides, there were a couple of casual inquiries.

Since then, I have been thinking and wondering about the pleasure and excitement of delving into randomly encountered math problems. It needs to be emphasized that math problems are not personal in nature, as are one's legal, medical, financial or psychological problems, and that are only to be shared with the so-called paid professionals. On the other hand, there is an element of universality about the nature of math as well as of math problems that are encountered even in high schools and colleges.

This problem was taken from a 100-year old classic textbook, ***Higher Algebra*** by Hall and Knight. It was discussed in the MAT 711 (**Survey of Mathematical Problems**) course, one of the three graduate courses designed specifically for the students doing master's with concentration in the **Teaching of Mathematics** (other concentrations are traditional - Pure, Applied and Statistics). A right textbook for this course has yet to be found. Therefore, the problems are gathered from all over to meet the diverse backgrounds of the students in this course. One of the objectives of the course that aligns with my teaching approach is to strengthen students' backgrounds in precalculus and calculus, as most likely, after graduation, they would be teaching in high schools and community colleges.

Mathematics is a dynamic activity in more than one way. The art of problem-solving is lost even to professionals in mathematics, if the mind is not sporadically engaged in stray math problems. At the age of 80 minus, I could not do this problem even with brute force, though I faintly remember having used inequality tricks years ago, then at my prime mathematically. By and large, problem solving in mathematics, like heavyweight boxing, is for the young!

In contrast to my student days in India of the 1950s, the US math curriculum has been getting softer over the last 50 years. In the name of not hurting the 'self-esteem' of the young, meritocracy is being thrown out. It is a reflection of the society at large, where people are consuming soft foods, soft drinks, playing on soft areas in gyms and parks, sitting on sofas for hours, and clicking away on smart gadgets for any information and indulgence. Textbooks in lower division math courses are inflated with too many unwanted hints and examples. Doing a problem means to mimic an example - not thinking. Consequently, independent performances are getting disastrous. On top of all this, currently, there are legislative and accrediting mandates on the institutions to turn out a large number of graduates in every degree program!

Coming back to the trail of problem solving, skills for tackling such math problems are different from the ones needed on a broad thesis or dissertation problems, where the mind remains logged in for weeks or even months. That requires a different mindset. Personally, a day in April 1961 is etched in my memory, when I went up to my instructor (in India then) for help on half a dozen old exam problems which I was unable to crack even after working on them for a few days. These days, the average time spent on a math problem by US students is hardly more than 10-40 minutes depending on the level of a math course. Well, my instructor was a renowned research mathematician, but he sweated to get even a single problem solved. Generally, there are two kinds of problem solving skill sets that are needed in mathematics, and they are not disjoint.

This reminds me of physicians who choose to work in the Emergency Rooms (ER) of the hospitals. They have very different mindsets, though they do get some medical training during residency. No two patients brought successively in an ER have any connection between their health problems. The physicians in the ER follow their guts while, at the same time, observing a medical protocol In the ER, there are no history charts available for the physicians to consult before examining the patients.

In India of the 1950s, students developed some problem-solving skills due to a comprehensive system of examination. There were no mathematics contests and competitions. However, the end-of-the-year final comprehensive exams were absolutely external - instructors having nothing to do with the selection of the exam problems and their grading. The preparation for the final exam was relatively brutal by the present US standards. For instance, I had sieved through several textbooks on differential equations to such an extent that I had coined the phrase, "I can smell a differential equation for its solution".

We all have 24 hours in a day, and, when they are gone, they are gone for everyone. Some distribute their time thinly on many topics, and some pour it into one pursuit, or most in between proportionately; perhaps, a Golden Mean. Nevertheless, a euphoric moment is met in the end when such a problem is solved. And, that moment can be carried forward into other walks of life too! Therefore, staying mathematically in shape is a boon.

Finally, all math graduate students should consider taking MAT 711 or MAT 712 (offered every two years) as an elective course. Personally, as a graduate student at Indiana University, Bloomington, I took 10 elective courses, which had nothing to do with my doctoral dissertation in **Partial Differential Equations**. However, they have been paying me off dividends for all these years.

October 25, 2018

PS Oct 30, 2018: one correct solution received!

COMMENTS

Dear Prof. Satish Bhatnagar, Namaste and thanks for the mail message and its contents.Yes, I urged my students for several years for more than a decade to go for three aspects of learning during their graduation (1) Developing a habit of self learning, learning on their own some new topic - without the help of teacher, (ii) Inculcating the culture of critical reading, reading between the lines by raising how, why, what if etc., and (iii) learning the art of problem solving.

I used to explain to them that whichever career they choose, more so if a teaching career is taken up, they will be at ease and will be comfortable if they developed the first one, they will require it very often to study many things on their own. Culture of critical reading will help them to raise many questions and by seeking their answer the concepts will be much more clear. And, the art of problem solving will indeed help them in handling day to day life problems as well - in addition to subject related problems. Researchers will not get success without it. However, the response of students and even junior / fresh colleagues too was poor - a fact noticed everywhere.

You might have observed that we in the ***Mathematics Student***, a publication of the IMS (Indian Mathematical Society), have started a Problem Section for students, teachers and for all readers - from the 2015 issue onward - thanks to Prof. Bruce Berndt and some other colleagues (in fact, well wishers of the IMS) who gave suggestions in this regard quoting ***Mathematics Students'*** earlier history and their assistance on the Editorial Board. Prof. B. Sury of the Indian Statistical Institute handles this section. It is three full years, and ***Mathematics Student*** is available on our website for open access, and a soft copy is sent to all life members individually as well.

However, you may verify that just five-six students take interest in it, and in the latest issue (you will soon receive) just one student/teacher has attempted to solve, successfully, few problems. I don't know the future, though we shall continue it. It appears that apathy towards problems is universal - though it is also a universal fact that right from primary to university level, in any subject and in any topic in that subject, take up any textbook, any chapter in it, any section in it and at the end of that section there will be graded set of exercises for readers/learners to solve, so that the concepts of that section gets firmly rooted in learners mind and the usual problem of "forgetting concepts or forgetting proofs of results or getting confused with proofs of several results", etc. can easily be tackled. But no one takes it that way, neither did we realize the importance when we were students!! It is like that only, since ages - with some exceptions perhaps!!! With regards, **J. R. Patadia.**

Thank you for sharing your ideas! Your text has provoked some thoughts, but you know, when one reads what other people write, this person mostly pays attention to what is tuned with his/her own thoughts and feelings. So, my comments can not be perfectly aligned with what you meant…

First, teaching Math 351, I noticed that it can be extremely effective to show students different approaches to the same problem. I mean not just "there are several ways to solve it", but the approach which can allow us to recast the problem and look at it from an absolutely different perspective. As an example: to attack the same problem you can give a combinatorial proof, an algebraic proof and/or the proof based on the principle of math induction. The other approach (it is probably a better thing, but I have just started looking into it and do not have a really good collection of such examples yet) would be to present the same problem (say, a pigeonhole principle) in different forms: as a geometric problem, as a combinatorics problem, etc. So, the envelopes (and the tools) change, but the essence is still the same.

The other thing is that at least one of the people who got your 27(a4 + b4 + c4) =/> (a + b + c)4 problem has started and hasn't finished it (to the best of my knowledge). Free time (the time to think) is an issue! We act more (or just function) than we think. You have probably heard about the famous (Russian? Ukrainian?) educator V. Sukhomlynsky (if you haven't heard, there is the link: https://en.wikipedia.org/wiki/Vasyl_Sukhomlynsky). One of the quotes from Sukhomlynsky is "The most important condition for the spiritual growth of a teacher is, above all, time — the teacher's free time. It is time to understand that the free time a teacher has, the more his spiritual world is devastated, the sooner that phase of his life will come when the teacher has nothing left to give to his students." I feel like some PTIs and GAs can (like me) suffer from rushing through the routine duties without enough time to focus on thinking one thought for more than 40 minutes (as you noticed, this is the time which the US students usually spend on their math homework). I wrote this last sentence and got scared of what I wrote. Here I will better stop before it all becomes too long. The very best wishes, **Viktoria**

Dear Sir, Many thanks for your email. I must say you possess an art of storytelling and your experiences in life plus everyday encounters with math problems continue to enlighten me. I actually found striking differences in the education system between now and 17 years ago when I was a student at Punjab. Perhaps, I may pen down some of my reflections that I experienced during my time at the university in England. I insisted on getting a degree from a western university this time and managed to complete a Master degree in just one year which usually takes around 2- 2 and a half years at any university in Pakistan these days.

Moreover, the quality of education being imparted at universities these days is an altogether different story. My friends who have just completed their MPhil degrees in Pakistan have told me their horrendous stories in terms of difficulties encountered in completing course work plus relationship with supervisors. They are university lecturers who completed their Master degree in 2000 (16 years of education) and now Higher Education of Pakistan (HEC) is funding their MPhil/ Phd degrees in order to build their research potential.

Consequently, Dr Husnine was really surprised that I paid a huge tuition fee to be educated at a foreign university and self funded my education. He doesn't know much about mathematics education discipline and its relevance in the teaching and learning of mathematics these days. He advised me to complete another Master degree in mathematics from abroad. Nevertheless, I am quite satisfied with the knowledge gained in this new discipline during the past year and excited to apply the knowledge of theory in practice.

On a separate note, I had a chat with an Assistant Professor in Mathematics here at a private university (University of Management and Technology, Lahore). His profile is https://ssc.umt.edu.pk/Academics/Faculty/Dr-Tabasam-Rashid.aspx His name is Tabasam Rashid and email is tabasam.rashid@gmail.com

He agreed to find sources regarding vice chancellors that have served University of the Punjab, Lahore, in the context of your research on the history of Punjab. Kindly contact him and provide him with the details of information you desire to gather. I hope that he will be able to successfully collaborate with you. Best wishes, **Sabahat Malik**

19. SLICING HISTORY OF MATHEMATICS

"If two mathematicians disagree on the solution of a mathematics problem, then at least one of them is wrong. However, if two historians agree completely on an historic issue, then one of them is taking a free ride on the shoulders of the other," thus, said I today during my lecture in a **Linear Algebra** (Math 330) class. That is the bottom line nature of these two disciplines. Mathematics is absolutely binary in the resolution of its problems; whereas, history is 'polynary' (my coinage).

This remark was made in the context of a problem of finding the eigenvalues of a matrix. It is reduced to solving a polynomial equation. Students knew the quadratic formula, but most of them did not know that the formulas existed for the cubic and biquadratic equations too. They were surprised to hear that all the proofs for solving equations of order greater than four had been wrong - it went on for two centuries. Then came Evariste Galois (1811-1832), a product of the French Revolution, who proved, at the age of 21, that such a formula does not exist!

Interjecting such snippets from the History of Mathematics (HoM) does not take me more than two-three minutes, but their impact on students' learning is long and deep. At this point, I made a brief pitch for the History of Mathematics (Math 314) course that I am scheduled to teach next semester (Spring-2019) - offered every two years. It is required for students majoring in Secondary Education Math who want to teach mathematics in high schools. Other students can take it as one of the elective courses, an option included in every degree program. The elective courses expand students' horizons in their twilight disciplines - a strength of the US college education.

This lecture being just a day before the Thanksgiving recess, I stressed upon the general importance of knowing the history of one's discipline, one's family, one's society and one's country. However, when it comes to teaching a course on HoM, no department of history would offer such a course, as most history faculty rarely go beyond pre-calculus. On the other hand, faculty in mathematics departments, in pursuit of their doctorates and research publications, get too far away from humanities and social studies. In order to teach any history course effectively, one must have a developed sense of history, which is evidenced by being able to connect the dots of an event in a manner that gives a unique interpretation. History is far beyond the memorization of dates of events and names of key people, just as mathematics is far beyond arithmetic calculations. History requires holistic thinking, and mathematics is all about deductive thinking.

At one time, I considered myself a history buff, but now I also claim (by Malcolm Gladwell's 10,000-hour dissertation in the O***utliers: The Story of Success***; 2008) to be an historian of contemporary events connected with the past. My book, ***Vectors in History, Volume I*** (2012) deals with history in general. Its 'historical' corollary, ***Darts on History of Mathematics, Volume I*** (2014) has 72 reflective and independent essays on varied aspects of HoM. The students of Math 314 have liked this book as a supplement to any traditional textbook on HoM.

I always look forward to teaching HoM courses, which are offered at UNLV both at the undergraduate and graduate levels. This is one of my intellectual passions as I am nearing the age of 80.

Nov. 21, 2018

COMMENTS

If two historians agree completely, they could both be wrong and probably are. **Paul Aizley**

Very interesting! I too am a fan of history in general and history of Mathematics in particular. When it became clear that I would be going to the USA for studies, I wrote to the USIS library in Mumbai if they could procure a suitable book on the history of the USA and they immediately sent me an excellent book on the subject. It helped me very much in that I could understand the motivations behind American policies when I was in the US. The last major book I read was "History of India after Gandhi" by Ramachandra Guha. As I came of age almost exactly as Gandhi died, I was almost a witness to all the events described in the book and was completely fascinated by the book. Thank you and as I said once before, *Keep 'em coming!* ***Arun Vaidya***

I certainly agree with all of that. I truly hope unlv appreciates the real gem they have in you to be able to bridge the two disciplines. Have a great Thanksgiving! **George Buch**

Satish, I find myself largely in agreement; thanks for sharing. I will add, though, that THIS historian went as far as multivariable calculus, and while he doesn't remember very much of it now, after 30 years, he took great interest in it back then and, unlike many in the humanities who almost proudly state that they are no good at math, would never do so himself. **Paul Werth**, UNLV History Professor

20. LEARNING-WHILE TEACHING!!

The teaching of History of Mathematics (HoM) courses, offered in mathematics departments and taught by math PhD faculty, provides pedagogical challenges of its own kind. Simply put, one cannot transfer literally and laterally anything from an experience of teaching a typical math course into that of a HoM course.

Also, I do not believe in a popular US cliche that if you want to learn some new material, then you go ahead and teach it. In the first place, who would take such an instructor knowingly? Not only has it trivialized the noble art of teaching, but has also harmed it in the universities, where the queen still remains Research (traditional). My test of teachability has been in having taken at least 3-4 courses beyond the course that is to be taught.

Ideally, an index of a well-taught course is a tangible return in the discovery of something new on the part of the instructor. Students are there to learn largely as a captive audience in courses. The pertinent question is the number of thought provoking moments for an instructor during the course or at the end of it. For me, this question is preeminent as while pushing the 80th year of my life, I continue to be a full time faculty member. Basically, I strive to reinvent myself professionally.

In this regard, the focus of this reflection is on the biweekly reports, one of the four criteria of evaluating the students in the HoM course (Math 314) that I am teaching during this semester. Today, I returned the biweekly report #3. As a data point, this course started with 25 students, but only one student has dropped it so far.

What are my expectations in biweekly reports? There are just three as follows. The Number 1 is critiquing the textbook material which is self-teachable. However, I do add some comments and paraphrase it in a few spots. In a typical math course, it is too much to expect from the students to do any self-study of hardcore topics in mathematics. The Number 2 is a critical commentary on a few reflections from a supplementary book, the **Darts on History of Mathematics**. It is a collection of 70 independent reflections on varied topics and themes in HoM. Again, in a math course, one textbook is more than enough! The Number 3 is a smorgasbord of impromptu comments which may come off the news of the day, from a random reading, or during class discussion etc. The whole idea is to make HoM come out alive. It does start jumping in the mind when HoM is approached as a subset of history in general. In most undergraduate math courses, the syllabus is so tight that during a lecture, even a slight deviation from the core material upsets the students. In one extreme scenario, the wholesome teaching is reduced to teaching to the class tests for a number of reasons.

During a HoM course, I do invite three guest speakers - faculty from other departments. They talk of their perceptions on HoM in the context of their specializations. Students are strongly encouraged to include the gist of guest presentations in their biweekly reports. Again, such interjections in a math course would be frowned upon by the students and the course coordinators.

Let me add that there is no structure in a biweekly report except that it has to be double-spaced and in 900 words +/- nine words. The word count impresses upon the young students the value of conciseness and compactness in non-fiction writings. Thus, students develop their talent in creative writing as a byproduct. I encourage the use of flowery language and the coinage of new terms and phrases especially drawn from mathematics. As a writer myself, I can already identify 4-5 budding writers in this class.

One may ask, if the biweekly reports are that useful, then why not to make them weekly? The answer is that the instructor's time is limited. For instance, it takes me about eight hours in 5-6 sittings to finish the reading of 24 reports. Unlike the solutions of the homework problems in a typical math course, these reports are all totally different from each other. Moreover, I am not a speed reader, and I do provide feedback in writing.

Finally, the reading of such biweekly reports is never boring - as such, my time is relatively overspent on them. But it is joyful to catch some interesting nuggets in some reports. In a typical math course, the monotonous job of grading a test can be charged to a teaching/graduate assistant. But a TA/GA is not helpful in a HoM course, as he/she would have little context while reading a biweekly report or grading a test. Viva HoM!

March 11, 2019

03/14/19 Postscript to **LEARNING-WHILE TEACHING!!**

After writing the above Reflection, I thought of getting brief feedback from my students in this course. They had already read a day ago what I had written. It was time to hear from them. And, I wanted their independent as well as anonymous feedback. Out of the 24 students in the class, 16 were present yesterday. In 5-7 minutes, I wanted them to give concise comments on what they clearly liked about the biweekly reports or did not like. Some students were lyrical about the reports and a few just put one thing down in three - four lines.

The following is a kind of tabulation of the salient points extracted out of their quick write-ups:

Not even a single student said that a biweekly report should not be a part of evaluation criteria. That validates how I feel so good about the reports.

.......keeps me organized
... like to reflect on topics and life... 4
...a .change of pace from a typical math course
....to recap and do additional research 4
....forces me to look back through notes and book 2
......help in connecting dots 2
....important exercise in combining information from different sources
.....fun.... likes its informal nature
.....pleasure to think critically
....get a better breakdown of reading
.....good and valid way to keep students engaged and note taking
.....are great and ...enjoyed them ...
....get carried away in writing...
.......helps me in class participation...
... .avoids skimming through the books....
...... express better in writing than in speaking....
.....writing is a struggle (being a math major)..
......900 words takes too much time - interferes with homework of other courses
.. Concerned about grammar
......Need more time for a perfect finish

…..Remark about word count - both +/- 7
……..Weekly reports and 400 words
……...Every three weeks and longer reports

COMMENTS

Once again, I think you are spot on. You describe precisely what a HoM should be and what you make it out to be. Not sure anyone else should even be allowed to teach it. Well, except maybe me if/when you do retire! Thanks for sharing! **George Buch**

I wish we had this class when I was a student. I've always enjoyed your lectures. You are definitely one of the best lecturers in my many math courses. You encouraged me to think, not just memorize. **Joe Mcdonald**

21. PROJECTS DEFINE HISTORY OF MATH

Since turning a non-fiction author at the 'ripe' age of seventy, I have become increasingly cognizant of the experiential aspects of teaching any course. More specifically, I must be able to describe what I have learnt from my students. At the same time, the students can take full ownership of having learnt something new by doing non-traditional exercises in my courses.

One of the five evaluation criteria in an History of Mathematics (HoM) course is a master project that carries 20% of the total grade. On Day One of the classes, students have no idea of what it entails. But its understanding evolves by the middle of the semester. They are encouraged to choose a project, though a sample of projects is provided. The whole idea behind projects is for the students to have hands-on experience with history. It is like a lab experience in a science course that helps in developing scientific thinking and scientific attitude in students. History projects develop interpersonal skills of the students as to how information is to be gathered from people known and unknown to them. The following is a list of projects undertaken by students during Spring-2019:

1. Korean math teacher
2. How technology has helped mathematicians in recent years
3. History of Twitter (with mathematical touches)
4. History of mathematics standards in US schools
5. International students' perspectives on math in the US
6. History of Math in Artificial Intelligence
7. Changes in technology in math over centuries
8. Math faculty in Foothill High School
9. Mathnasium, its origin and how it has progressed
10. Women in mathematics PhD
11. Math faculty at Coronado High School since 2000.
12. Mathematics of the Native Americans
13. History of Planck's Constant
14. Comparative history of (higher) education in the US and China
15. History of women in mathematics for the last two centuries.
16. History of computers related to math.

17. Why people study mathematics; history and emotions behind it
18. Math faculty at my high school alma mater
19. Evangelo Yfantis, UNLV Professor of Computer Science
20. History of the impact of math on literature
21. Lesser known facts about Einstein, Newton, and Gauss
22. Katie Bouman - her math background, algorithm, and information
23. History of mathematics in computer science
24. Evidence of Mathematics in Ancient Structures

Here are a few observations drawn from these projects:

1. Students were not tied down to a topic of their first choice. For whatever reason, if they changed their topic, it was also a learning experience in finding facts in history. I am proud to say these students will have the best understanding between facts in mathematics and facts in history.

2. The project has to be written up in 1000 words +/- 10 words. That instills the value of mathematical compactness and conciseness in writing too - besides edits and revisions. Pursuit of mathematics does make people slow readers, and I am one of them. Turning in of the grades within a day or two of the final exam does not allow the reading of lengthier projects.

3. The range of topics is wide - like, from a story of Planck's Constant to the story of Katie Bouman.

4. For the benefit of the entire class, each student presented a summary of his/her project in 3-4 minute - a week before the final week of instruction. The final draft was due in the last week. Presentations were duly timed to impress upon the students the value of a good oral presentation.

5. I had told the students that any project drawn from any online sources will not carry the same weight in evaluation in comparison with a project where one goes out to interview or gets information by direct communications. There were no restrictions on places and people.

6. Some students realized that communication skill is needed for getting a stranger to talk with and solicit specific information. In some cases, I did speak on behalf of the students.

This course is offered every two years. It is required for the majors in Secondary Mathematics Education, but not for math majors. I always look forward to teaching it.

May 09, 2019

22. UTILITY OF A COLLEGE COURSE

"What use have you found in this History of Mathematics (HoM) course?", asked Josh Culver from Aaron Harris, who had taken my History of Mathematics (MAT 714) course in the Spring 2007. Josh Culver, a math graduate student, is currently enrolled in this course. This Q-A exchange took place yesterday. Aaron Harris, after finishing PhD in Mathematics Education, a doctoral program housed in the UNLV's College of Education, is now a mathematics professor in the College of Southern Nevada. For years, he taught math in a local high school. I had Invited him to give a guest lecture on what HoM meant to him.

At the spur of the moment, Asron's answer was perfect when he said that the course was about the wider awareness of mathematics whose abstract nature is humanized after knowing some HoM. He likened this question with a course in English composition that he took years ago in which various technicalities of English grammar rules were emphasized. He admitted that he had no ideas as far as any use of those rules was concerned. He summed it up by saying that all courses of study are about training the mind for different modes of thinking. I would further add that taking different college courses is like going to a modern gym, where different exercise machines help in the development of different muscles of the body.

The utilitarian aspects of any intellectual activity is a perennial question. Both high school and college students raise such questions in the classroom sometimes to the point of instructors' annoyance. Such an attitude is more prevalent in societies and families where education is free, compulsory, or easily accessible - like in the present USA. As a matter of fact, it goes back to eons of time in any society. We just read about it in ancient Greeks in reference to Plato. This course really brings out thinking in a very sharp manner. My attempt goes wider and deeper by integrating humanities with mathematics. Here are some of the salient points to be taken away from any HoM course:

1. The facts in mathematics, whether in its definitions, postulates and axioms are crystal clear.

2. In history, the facts are subjective and change with individuals. There are no building blocks of facts in history- like point, line, and angle etc. are in geometry.
3. Hard core mathematics is binary in nature - a solution of a problem or a proof a theorem is right or wrong - nothing like 90% correct!
4. History includes various valid points of views on any issue.
5. The proofs of theorems and solutions of math problems do not change with time.
6. In history, revision is constantly taking place, as new facts are being discovered or a new political system takes over.
7. Math textbooks before 1970 did not have even a line about mathematicians. Afterwards, they were revised to include some footnotes to humanize math.

Finally, I urged my students to add some thoughts to the above list, and include them as a part of the next weekly report

October 02, 2019

23. KNOWING THY HISTORY!

During the last three offerings of the **History of Mathematics** (Math314) course, students essentially chose their individual master projects. Nearly half of them did it in a timely manner, but the other half either followed my suggestions late or just put together half- hearted ones. Moreover, by the time they were done choosing their topics, hardly 4-6 weeks were left for them to work on them. Thus, full justice could not be given to this important aspect of the course.

The whole idea of doing a 'history' project in this course is to instill **history thinking** akin to taking a science course with a lab, as only then, foundations of **scientific thinking** are laid in. Basically, history is about observing, discriminating and compiling facts, which remain subjective and so does their interpretation. That sets history miles apart from mathematics - a useful lesson to be learnt.

There is one major boundary condition in doing a project in history - that it is not to be totally done by Google searches and library visits. In other words, footwork is encouraged in the sense of going out and meeting people who are possible resources. The tools of oral history are encouraged. A guest lecture on **February 10** will speak on the **Importance of Oral History** and explain how to approach people and conduct interviews.

This year, I decided to assign the same topic to the entire class in the second week of the semester. That gives the students ample time to think and work slowly but steadily. T**he topic is to bring out the life and time of any one of one's great grandparents**. Everyone has four grandparents - taking into consideration both sides - mother's and father's, it implies everyone has eight great grandparents.

1. The bottom line for this choice is that of all histories around, knowing one's family history is the most important one. It is just like - planning for the future depends on the path undertaken in the past - from yesterday to down.
2. This research will take the students to only 60 - 80 years back in time, maybe in different countries - yet close to the present.

3. In the context of the chosen ancestor, it is worthwhile to find out his/her social and political involvement.
4. The schools and colleges he/she attended.
5. Information about math curriculum, math textbooks, names of schools, and teachers at any level.
6. Any artifact used by this ancestor, or any piece of legacy.

The following lessons are to be drawn:

1. Facts in mathematics are unassailable. In history, they are subjective, thus changeable!
2. This exercise provides a reality check to the timeline of any history. Often, the pages of history are turned quickly covering decades and even centuries. Yes, relatively, little is known as the past is dug deeper, but that does not mean nothing happened on any given day in the past!
3. History brings out the fact that Time is a great destroyer and a great filter. It applies to both tangible and intangible objects.
4. Mathematics is the worst victim of the mighty Time!

The final draft of the project report, in 1200-1300 (double space) words, is due on **Monday, 04/20/20**.

January 24, 2020

24. FRACTIONS - FRACTALS - FRICTIONS!?

Today, while adding my commentary on a kind of a brief 'History of Fractions' chapter, and in order to personalize fractions too, I asked my students to recall their first exposure to fractions in elementary schools and their subsequent dealings with fractions in middle and high schools. For this exercise, they had ten minutes to put down their thoughts on one page only.

There is no point in studying unsystematic symbolism of fractions as used in ancient Egypt, which were modified as the ideas **supposedly** moved to Greece and Babilonia eventually. Showing a map of the region, I raised questions as to what really happened as ideas must have moved in all the four directions from Egypt, though at different rates - to the west in present-day Sudan, south in Ethiopia, east in Israel and north in Italy. **All History of Mathematics (HoM) textbooks talk only of Greece in the northeast of Egypt**. Even the closest region, around the present Israel, is completely blanked out mathematically despite the evidence of great monuments built there 3000 years ago.

I took this opportunity to emphasize that **research in any field is always driven by financial, political, academic and cultural considerations.** For instance, yesterday, President Trump announced in his State of the Union Address an allocation of $25 million for prenatal research that may save babies who are born even after four weeks in pregnancy. Who is there to fund research in the ancient HoM? That is why to the best of my knowledge, there has not been even an iota of systematic addition to HoM in the last 100 years.

Anyway, below are given a gist of students' encounters with fractions as they recalled in a time bound setting. The numbers in parentheses are the number of students who gave that particular comment or a close one. All thirty-three students in the class turned in their write-ups. A couple of them did not say a significant thing, but a few had two observations:

1. Struggle with additions and subtractions of fractions in elementary schools. (6)

2. Multiplication of fractions. First, multiply numerators and denominators and then simplify. Second, first cancellation and then simplify fractions.
3. Liked fractions in high school and college. (4)
4. Visual introduction of fractions - a piece of a pizza. (12)
5. Overwhelming in the beginning, but fine after practice for a few years. (8)
6, Written fraction and yet being 'whole' was confusing.
7. A religious connotation of good and bad with numerator and denominator. (2)
8. Relating with percentages was fun.
9. 'Fraction' storage in computer and rounding errors.
10. Slope of a line, 'rise over run', seen as a fraction.
11. Fractions in everyday phrases.
12. Fractions understood in block formations.

As far as my encounter with fractions is concerned, nothing stands out vividly - no pictures, no models; nothing. Partly, it is due to a gap of at least 65 years - some memories are erased. Even the elementary education in colonized India (before 1947) was a privilege for the urban kids. The bottom line being that it is easy to subjugate an uneducated population. Also, by educating people in a system of education imposed by the colonizers, challenges to the authorities are minimized. Thus, there being little freedom to think, we never questioned our teachers. However, in the context of fractions, once the techniques of finding the Least Common Multiple and Greatest Common Divisor/Factor of natural numbers were understood, I think factions were tamed! That is all I can recall.

At UNLV, Math 314 (HoM) is required for the students majoring in Secondary Education Mathematics. My hope is that this exercise will register in students' minds so that when they eventually become school teachers their students would sail through **Fractions** without **Frictions** - and not be buffeted by any coastal **Fractals**!

February 05, 2020

COMMENTS

Fractions have occupied my mind for many years. I have learned with great mental effort that there are more than one interpretation/meaning/ application of this concept. Almost all textbooks have inadequate/ incomplete definitions. No wonder students (and later in their adult lives) 'hate' fractions. I once was telling a friend of mine the horror story of how my students add 1/2 + 3/4 = (1+3) / (2+4). To my surprise/distress, the friend said "Isn't this the way to add fractions? And, this person was a CPA and tax accountant for many years!!! **Raja**

Love the idea of students writing about fractions. I will try this in my class as well. Thank you!!
Renuka Prakash

25. SQUARING JOY IN PROJECTS

Any History of Mathematics (HoM) course, being relatively more flexible than a typical math course, gives an instructor tremendous freedom in terms of the choice of topics, emphasis and its evaluation. Since 2007 when I regularly started teaching HoM courses both at the graduate and undergraduate levels, I have been emphasizing the importance of individual projects. The objective is for the students to understand what the facts in history are, and how they come about, and how they appear in textbooks and mainstreams of ideas etc. That is very different from a mathematics content course. Moreover, the students get a strong feel for the timeline in the HoM, subjectivity in it, importance of oral history, and socio-political culture of an era.

This year, in the HoM course, (Math 314), I decided on a project that was personal and emotional to each student besides the above considerations. It was explained that everyone has two biological parents, four biological grandparents, and eight biological great grandparents, and sixteen great great grandparents, and so on. Choose any one of the sixteen great great grandparents. In 1200 words, bring out a story of his/her life, and spice it up with mathematics in terms of the schools/colleges attended, teachers, books and curriculum etc. The students had twelve weeks to work on it. A guest lecture on the importance of Oral History helped the students in collecting information.

In a typical math course, there is so much monotony that there is nothing new and refreshing while teaching, discussing a problem or grading a quiz or test. In contrast, I looked forward to reading these projects. I finished reading all 33 master projects over a week. With my slow reading speed and the surprising nature of the material of each project, I could read only 4 -5 projects in one sitting. All along, I was also copy-pasting highlights from each report, evaluating them, scoring them, and entering the scores in a spreadsheet.

The reading experience was like a series of rides on a roller coaster, but with a difference. In a physical roller coaster, one experiences the rise and fall once or twice, but in the reading of these projects, it seemed like an unending sequence of fluctuations that I felt groggy a few times!.

In the previous offerings of this course, master projects have been integral to the instruction, but their formats were different - more clinical than personal. Just for trying something different this year, when this project struck my mind, I had no idea of its shaping up. More than half of the students had their roots in other countries. The reading of the stories of their great great grandparents was kind of refreshing.

Personal enrichment is a hallmark of learning in teaching a course. It applies to student researchers and to me as their evaluator. This project met this goal. Some of the students are destined to become writers and some as intellectual miners.

May 07, 2020

SECTION II

HUMANISTIC SLICES

26. MAKING AND UNMAKING OF BENJAMIN BANNEKER

This is neither my typical reflection nor an article. This is extracted from a paper that I presented at the Joint Mathematics Meeting held in Seattle in Jan 2016. The points that I raised there are worth including in this book, and hence this write-up. Some of the ideas had surfaced up after reading the book, ***The Life Benjamin Banneker*** by Silvio Bedini (1999). Benjamin Benneker (1731 - 1803) lived through the battles waged after the 1776 Declaration of Independence.

Open Questions: 1. The fire that destroyed Benjamin Banneker's (BB) house only two days after his death, and that too while his memorial service was going on – remains a mystery.

2. A person, who published six almanacs, must have kept some personal diaries too. It reminds me of Solomon Northup, who wrote in 1853 his story of life of 12 years as a slave. Nothing is known of BB's writing that shed light on his personal life.

3. Above all, BB's never marrying does not ring right. His ties with women are mysteries too. He had three sisters, and at least one of them must have taken a keen interest in seeing their only brother married.

Coordinates of Benjamin Banneker's Life

1. BB's grandmother, a white Welch woman, came to the American colonies as indentured labor. As a sidebar, it may be noted that the parents of Benjamin Franklin, a US Founding father, had emigrated from England. BB's grandmother went through the hardships of a slave's first hand. Later on, she freed her half a dozen slaves and married one of them. She gave birth to only one child, a daughter - who repeated her mother's story by marrying one of her freed slaves and giving birth to BB besides three daughters.

2. After middle school, all-homeschooling built BB's intellect. He grew up on the classics of English civilization. The White farm neighbors provided

support for BB's self-study of science and astronomy. BB may be the first polymath of a young nation, USA.

3. BB's quiet demeanor and his lifelong cordial ties with his White neighbors built his public persona locally. Later on, some of his White acquaintances yielded influence in the public life of the colonies.

4. By modern standards, BB was an applied mathematician. He understood math which was needed in the computation of sunrises, sunsets, phases of the moon, tides, and other statistical information and related topics. Astronomical or meteorological information usually include future positions of celestial objects, star magnitudes, and culmination dates of constellations.

5. BB was far more than a handy man – a real inventor. Historically, more inventions in the US are born in the home garages than in fancy government labs.

Galois (1811-1832) and Cauchy (1789-1857) born before Banneker means that modern mathematics was hardly on the horizon in Europe. Moreover, all the British colonies were kept in relative 'darkness'. Thus, the American 'renaissance' did start around 1776. A lesson of history is that without political freedom, there is no blooming and blossoming of the minds.

Why my interest in Black History? It runs parallel to the contributions of the Hindu scholars, which have been denied/diluted due to the colonial conditions in India that started from the 12th century.

February, 2016

27. EXPANDING MATHEMATICAL HORIZONS

(A Publicity Pitch)

"*In mathematical jargon, the objectives of this session are to explore the necessary and sufficient conditions for the development and flourishing of mathematics. From a cursory examination of history, it is clear that these conditions are functions of at least three factors; namely, political systems, educational models, and organized religions. These are structured institutions and are reflected in the growth of mathematics as a super organized discipline.*

At any point in history, if the entire world is surveyed mathematically, then one finds a few green spots (countries) where math thrives (homes of the Fields Medalists and Abel Prize winners, etc.). There are brown spots where fundamental math research is not on the horizon. Then, one sees swaths of barren spots, where math is stressing to sprout. Thus, papers are invited where the focus may be on an individual, society, institution, or ideology in the present or at any time and place in the past for the validation of these objectives. This session cuts across the traditional boundaries of history, politics, psychology, sociology, science and religions."

This is a posted description of a session on the **History of Mathematics** that I am organizing during (April 18-19, 2015) a meeting of the Western Section of the American Mathematical Society to be held on the campus of UNLV. Here is a link for further details: http://www.ams.org/meetings/sectional/2218_program.html

The reason it is being circulated is due to the breadth of its disciplinary scope. A curious layman, an undergraduate, or a graduate student can contribute. Thus, it is not limited to mathematicians or historians. Lately, my personal focus in the history of mathematics is on the impact which the organized religions, political ideologies, educational curricula, and models of government have on mathematics.

A statistician friend of 50+ years told me that if there is data, then he could analyse it. However, there is no such data available, say, on the number of top–notch mathematicians who are adherents of major religions – like, Hinduism, Judaism, Buddhism, Christianity, Islam and Sikhism. That means I have a challenging task of creating some kind of data too. And, that is where I decided to seek out the help of my various Reflection readers. To get it going, a top-notch mathematician may be a winner of any one international mathematics award like – Fields and Chern medals or. Abel, Polya, Nevanlinna, Leelawati, Gauss prizes, etc.

However, these honors set a high and unrealistic benchmark. For the sake of this session, research mathematicians of international fame, working in famous universities (Tier One in the USA), or institutes are to be included. Mere PhD in mathematics or full professorship is not enough. It does get subjective; but that is the nature of the history of anything. It would be fun to see all kinds of names.

Yes, forward this e-mail to any friend or colleague who may be interested. My hope is to have data that may speak clearly that certain organized religions, societies and systems are conducive to mathematical thinking, and so are some forms of educational, governmental and economic models. A bottom line in my research is that science and mathematics are the products of organized societies and they tend to flourish both in times of peace and of war. Also, Google the Nobel Prize winners since 1905, and use your creativity to extrapolate it in terms of mathematics.

Thanks in advance for helping me out – the sooner the better. In turn, your horizons will surely be extended.

Jan 20, 2015

28. PUTTING THE TWO TOGETHER!

On Jan 21, I wrote a mathematical reflection, **EXPANDING NEW MATHEMATICAL HORIZONS** inviting participation in the history of mathematics session to be held in the AMS sectional meeting on April 18-19, 2015. Here is an extract:

"***In mathematical jargon, the objectives of this session are to explore the necessary and sufficient conditions for the development and flourishing of mathematics. From a cursory examination of history, it is clear that these conditions are functions of at least three factors; namely, political systems, educational models, and organized religions. These are structured institutions and are reflected in the growth of mathematics as a super organized discipline.***

"***At any point of time in history, if the entire world is surveyed mathematically, then one finds a few green spots (countries) where math thrives (homes of the Fields Medalists and Abel Prize winners, etc.). There are brown spots where fundamental math research is not on the horizon. Then, one sees swaths of barren spots, where math is stressing to sprout. Thus, papers are invited where the focus may be on an individual, society, institution, or ideology in the present or at any time and place in the past for the validation of these objectives. This session cuts across the traditional boundaries of history, politics, psychology, sociology, science and religions.***"

"This is a posted description of a session on the **History of Mathematics** that I am organizing during (April 18-19, 2015) a meeting of the western section of the American Mathematical society to be held on the campus of UNLV. Here is a link for further details: http://www.ams.org/meetings/sectional/2218_program.html

"The reason it is being circulated is due to the breadth of its disciplinary scope. A curious layman, an undergraduate, or a graduate student can contribute. Thus, it is not limited to mathematicians or historians. Lately, my personal focus in the history of mathematics is on the impact of organized religions, political ideologies, educational curricula, and models of government have had on mathematics."

On Jan 31, 2015, the Yahoo News headlined, **IRAQI LIBRARIES RANSACKED BY ISLAMIC STATE GROUP IN MOSUL.** Here are some extracts: (Bolding is mine)

BAGHDAD (AP) — When Islamic State group militants invaded the Central Library of Mosul earlier this month, they were on a mission to destroy a familiar enemy: other people's ideas. Residents say the extremists smashed the locks that had protected **the biggest repository of learning in the northern Iraq** town, and loaded around 2,000 books — including children's stories, poetry, philosophy and tomes on sports, health, culture and science — into six pickup trucks. They left only Islamic texts. The rest? "These books promote infidelity and call for disobeying Allah. So they will be burned," ……

Since the Islamic State group seized a third of Iraq and neighboring Syria, they have sought to purge society of everything that doesn't conform to their violent interpretation of Islam. They already have destroyed many archeological relics, deeming them pagan, and even Islamic sites considered idolatrous. Increasingly books are in the firing line.

Mosul, the biggest city in the Islamic State group's self-declared caliphate, boasts a relatively educated, diverse population that seeks to preserve its heritage sites and libraries. In the chaos that followed the U.S.-led invasion of 2003 that toppled Saddam Hussein, residents near the Central Library hid some of its centuries-old manuscripts in their own homes to prevent their theft or destruction by looters. But this time, the Islamic State group has made the **penalty for such actions death**. Presumed destroyed are the Central Library's collection of Iraqi newspapers dating to the early 20th century, maps and books from the Ottoman Empire and book collections contributed by around 100 of Mosul's establishment families.

Days after the Central Library's ransacking, militants broke into the University of Mosul's library. They made a bonfire out of hundreds of books on science and culture, destroying them in front of students.

A University of Mosul history professor, who spoke on condition of anonymity on account of his fear of the Islamic State group, said the extremists started wrecking the **collections of other public libraries last month**. He reported particularly heavy damage to the archives of a Sunni Muslim library, the library of the 265-year-old Latin Church and Monastery of the Dominican Fathers and the Mosul Museum Library with works dating back to 5000 BC.

Here are my thoughts for pursuits: It is a kind of contrapositive format to the necessary and sufficient conditions for the positive development of mathematics are the social and political conditions, which are negative, and thus inimical to the growth of mathematics. The region controlled by ISIS is bigger than many countries of the world. It is safe to conjecture that mathematics and science are not going to take any roots in the near future.

Therefore, it is a perfect topic for someone to write a paper for this conference. That is putting the two together!

Feb 09, 2015

COMMENTS

Satish, I wish you well in the AMS sectional meeting and the conference. But it is perfectly clear that ISIS wishes to take people back to the Dark Ages and the first thing that has to go is knowledge for it is the most dangerous thing to their ideology. Kind of like the gang of outlaws that holds a town hostage in the old west only gargantuan in scale now. Very sad and very scary. **Brian**

Dr. B., Thanks for sending this. I hadn't seen it. We'll have to see if Obama brings it up at the next prayer breakfast and tries to excuse ISIS by saying Christians did the same thing......**ONN**

ISIS Burns 8000 Rare Books and Manuscripts in Mosul

The Fiscal Times By Riyadh Mohammed
Feb 23, 2015

While the world was watching the Academy Awards ceremony, the people of Mosul were watching a different show. They were horrified to see ISIS members burn the Mosul public library. Among the many thousands of books it housed, more than 8,000 rare old books and manuscripts were burned. "ISIS militants bombed the Mosul Public Library. They used improvised explosive devices," said Ghanim al-Ta'an, the director of the library. Notables in Mosul tried to persuade ISIS members to spare the library, but they failed.

The former assistant director of the library Qusai All Faraj said that the Mosul Public Library was established in 1921, the same year that saw the birth of modern Iraq. Among its lost collections were manuscripts from the eighteenth century, Syriac books printed in Iraq's first printing house in the nineteenth century, books from the Ottoman era, Iraqi newspapers from the early twentieth century and some old antiques like an astrolabe and sand glass used by ancient Arabs. The library had hosted the personal libraries of more than 100 notable families from Mosul over the last century.

During the US led invasion of Iraq in 2003, the library was looted and destroyed by mobs. However, the people living nearby managed to save most of its collections and rich families bought back the stolen books and they were returned to the library, All Faraj added."900 years ago, the books of the Arab philosopher Averroes were collected before his eyes...and burned. One of his students started crying while witnessing the burning. Averroes told him... the ideas have wings...but I cry today over our situation," said Rayan al-Hadidi, an activist and a blogger from Mosul. Al-Hadidi said that a state of anger and sorrow are dominating Mosul now. Even the library's website was suspended.

"What a pity! We used to go to the library in the 1970s. It was one of the greatest landmarks of Mosul. I still remember the special pieces of paper where the books' names were listed alphabetically," said Akil Kata who left Mosul to exile years ago. On the same day the library was destroyed, ISIS abolished another old church in Mosul: the church of Mary the Virgin. The Mosul University Theater was burned as well, according to eyewitnesses. In al-Anbar province, Western Iraq, the ISIS campaign of burning books has managed to destroy 100,000 titles, according to local officials. Last December, ISIS burned Mosul University's central library.

Iraq, the cradle of civilization, the birthplace of agriculture and writing and the home of the Sumerian, Akkadian, Assyrian, Babylonian and Arab civilizations had never witnessed such an assault on its rich cultural heritage since the Mongol era in the Middle Ages. Last week, a debate in Washington and Baghdad became heated over when, how and who will liberate Mosul. A plan was announced to liberate the city in April or May by more than 20,000 US trained Iraqi soldiers. Either way, and supposing everything will go well and ISIS will be defeated easily which is never the case in reality, that means the people of Mosul will still have to wait for another two to three months. Until then, Mosul will probably have not a single sign of its rich history left standing.

29. MEASURES OF GREATNESS...!

A course on the History of Mathematics (HOM) can be the most interesting course for its instructors as well as the students. The class discussion can take off tangentially in any direction whether from a student's remark or instructor's insight. Four weeks ago, I asked my students in the undergraduate course on HOM (Math 314) to compare two nations in terms of 'greatness' – no other specifics. Americans are always comparing, ranking, and scoring all kinds of tangible and intangible attributes in life. I gave them only 5-7 minutes to put down one concise measure each as to when Nation A is 'greater' than Nation B.

Here are the raw and unedited responses from a class of 20 students. Some of them were absent that day and a few had no ideas or just did not turn in their responses. It is a class of mostly juniors and seniors, all are at least 20 years of age. Most are majoring in high school mathematics teaching, where this is a required course. However, there are quite a few majors from mathematics, computer science and other areas, where this course is not required. Here follows the list in not particular order:

1. Nation A is greater than a Nation B, if Nation A has more knowledge of the sciences.

2. A nation's greatness is determined by how they take care of their weak and/poor, young and old.

3. By the impact it has on other cultures or civilizations.

4. We can define that a nation is greater than another on its preparedness on economic issues.

5. A great nation is one in which we have freedom to do as they wish without stepping on those who cannot.

6. A nation is great if it allows the sciences to flourish.

7. A great nation is one that takes care of its citizens.

8. Nation A is better than Nation B if Nation A does more for the advancement, betterment of the lifestyles of its people, not at the expense of any one.

9. The overall contribution to human knowledge and welfare.

10. I think education defines a nation's worth.

11. A nation is greater than another nation if a nation is more economically stable.

12. They understand the importance of certain aspects –what works and what doesn't.

13. Nation A is greater that Nation B if A owns the people of B, be it through resources, slavery, or idealism/culture. Thus, a nation's greatness is the degree to which all of its peers are influenced inevitably, the greatness of a nation is observable by the probability a nation is referenced in any general history book, and regardless of the time or place, that book is found.

The students' responses are essentially spontaneous – from the top of their heads, as I had wanted. There was little build-up in the background. However, I have pondered it over for years and here are just a few of my criteria, but without any explanation.

Mathematics is the index of a nation's greatness.

Nation A is greater than Nation B, if the women of Nation B go after the men of Nation A, or equivalently, men of Nation B do not court and woo the beautiful women of Nation A.

The net inflow of people who would move into a nation, if the immigration doors are hypothetically left open for a day or two for any foreigners to enter.

It is a fun exercise with several corollaries to contemplate. Feel free to share it with anyone.

March 21, 2015

COMMENTS

Hi Satish: I will give the same question to my class on the history of science. I will let you know their response. Stay tuned!! **Alok Kumar**

It's really fun to read the responses. I love it. I miss being a teacher! **Shaloo**

I wrote: Opportunities/Environment for realizing one's dream is another measure of a nation that you can be what you want to be. Here the US fits perfectly. Explore various programs that Atlanta offers in abundance of all kinds, and start one next year. Yes, it will take one year of planning!

Hi Satish Ji, Very interesting. Why so many different criteria? I believe, often, human beings tend to justify what they personally think is the right way to do it. In this specific case, people would consider how to justify the greatness of the nation (they think) they belong to in the light of their personal knowledge/ comfort/ area of expertise/ experience. Sometimes, some agencies/organizations tend to develop quantitative/qualitative algorithms. This may get accepted by others and utilized by the rest of the world. However, one cannot exclude the possible influence of personal/ organizational sphere of knowledge or agenda. One thing is certain, all people think and they do a good job at it. Regards, **Kishore**

Dr. Bhatnagar-ji: Thanks for sharing these interesting results of your impromptu survey of college kids. I guess each one has a different world view. I believe in the statement below, which you wrote: The net inflow of people who would move into a nation, if the immigration doors are hypothetically left open for a day or two for any foreigners to enter. This, I do believe, is why the U.S. is greater than any other nation!

Another important one would be, if the index of happiness (measured objectively) of a nation is higher than that of another. After all, what is greatness, if there is no or little happiness! Thanks for letting me share my thoughts with you. Regards, **Mohandas Bhat**

Very interesting. I'm surprised there is no mention of military power. Also, notice how most did not really attempt to detail how one nation might be considered greater than another. I think our younger generations have been influenced by so much "fairness" (e.g., not keeping score in Little League Baseball so losers don't feel bad) they have trouble with the concepts of "better" or "best." I think your immigration definition nails it.

My take: Nation A is greater than nation B if the citizens of nation A enjoy more individual freedom and less oppression than the citizens of nation B. And that would explain the immigration corollary. **Charlie**

Interesting-I can see some of your academic influence on their minds (a good thing). I've always felt a class takes on a bit of the personality of their instructor just as a team takes on the personality of their coach. I have never considered your second statement about men and women (as well as the contrapositive)...thought provoking. Nation A is greater than nation B when the quality of character of the median person in nation A is significantly better than that in nation B. Character measures may include a variety of variables such as grit/determination, attitude, beliefs, work ethic, education (especially mathematical!), piousness, kindness, creativity, and service. What do you think? **Aaron Harris**

I wrote: In a similar vein, I often say that the greatness of a nation can be seen even in its ordinary citizens.

30. VON NEUMANN SPIKE!

The moment I noted '***Von Neumann Spike***' in one of several slides during a colloquium talk on the '***Ultrafast Shock Compression Experiments to Rapidly Test Extreme Condition Materials Predictions***', my attention was diverted away from the lecture. The speaker, a chemical physicist, was from the world famous Lawrence Livermore National Lab. ***'The recent results from ultrafast tabletop laser compression experiments on fluids, polymers, and high energy density organic molecules including single crystals.'*** from the abstract of the talk pulled me to the seminar room of the **Physics and Astronomy Department**. In most US universities, Friday afternoon is a time to arm wrestle with the cutting edge ideas in the seminars, which often continue over drinks in bars and pubs situated near the college campus.

For me, being surrounded by experts from different fields is no more awe-inspiring. I simply go there to get random stimulation to my intellect from the 'sights, sounds and smells' of different trends and the latest in research. It fertilizes the pores of my mind. The various units and nanoscales, the speaker was talking about, floated over my head. He mentioned a covered lab facility, which is as large as the area of four football fields - nearly 250,000 Sq. ft. The type of research that goes on there alone is beyond the scope of 90% of the countries. This is what makes me wonder at the disparity between the US and those countries of the world. All other disparities are its corollaries. Historically, the US is only 225 years old - younger than most countries in the world!

Neither I wanted to interrupt the speaker over my question on ***Von Neumann Spike***, nor did I want to wait for the Q-A session at the end. An iphone being with me, I just Googled it in my lap. It turned out to be named after 'the' John Von Neumann (1903-1957), a polymath, who did fundamental research in a dozen of mathematical areas - including, Game Theory, Linear programming, Operator theory and Artificial Intelligence. Born in Hungary and educated in Switzerland early on, Von Neumann published one major paper a month! He is one of the first four permanent members of Princeton's Institute of Advanced Studies – the remaining three being Einstein, Gödel and Herman Weyle.

During World War II, Von Neumann was totally immersed in the Manhattan Project and was consulted by the national labs. The principle of ***Von Neumann Spike*** came out of his 1941 research in detonation theory. Stanislaw Ulam (1909-1984), the Polish top mathematician in the Manhattan Project, was Von Neumann's closest friend. As the Nazis were rounding up the Jews in Poland and Hungary and herding them off to the concentration camps, Von Neumann and Ulam helped several Jew scientists and mathematicians escape and move them to the US. Most of them were immediately put to work on the Manhattan Project.

This transfer of the brain paid off enormously, as these mathematicians and scientists researched their respective problems with incredible frenzy. Defeating Hitler by producing the first Atom Bomb became the 'personal' goal of each one of them. The development of Atom Bomb is the most amazing story of the ages- far bigger than any fictional one.

January 29, 2016

COMMENTS

That's very interesting, Dr. Bhatnagar. Von Neuman has so many papers that he literally shows up everywhere. **Drew**

Dear Satish: Thank you for sending me your thoughts! I remember the Von Neumann spike mentioned in Joe Zaug's talk and did realize it as yet another contribution from Von Neumann who was a genius polymath. It was nice to see you there last Friday! All the best, **Michael Pravica**

31. BRICKS AND BOUQUETS ON THE *DARTS*

For me, the writing of a book and enjoying it is like fathering a child and loving it. Most parents (modulo a set of measure zero) do love their children, but very few people write books, and much less are proud of them. The ***Darts on History of Mathematics*** is an outgrowth of my miscellaneous articles and reflections on the History of Mathematics (HoM) that I have been writing since the 1980s. Most of these reflections were prompted by my teaching of the undergraduate and graduate courses on HoM during the last 15-20 years, and attending the HoM conferences and sessions in the US and abroad.

After the publication of the ***Darts*** (for abbreviation) in 2014, I casually used it as a supplementary book in a HoM course. Because of its no-sequential style in terms of topics, eras, cultures, and regions, it is not meant to be used as a textbook. On the other hand, there is nothing like the ***Darts*** as far as its use as a supplementary book is concerned. That has been my opinion, which may have an author's bias too. Therefore, seeking validation from the students was essential.

On May 03, 2017, the last day of the spring semester, I asked my 26 (out of 27) students of HoM course (Math 314) to use one side of the sheet for the evaluation of the textbook. ***Math Through Ages*** by Berlinghoff and Gouvea (2015) and the other side for the evaluation of the ***Darts***. On the sheet, the spaces were marked out for the writing on just two ***Strengths***, and only one glaring **Weakness** of the two books. I stayed out of the classroom until everyone was finished. The names on the sheet were optional.

Here is the summation on the strengths and weaknesses of the ***Darts*** from the students' perspective. The numbers in the parentheses are the numbers of students who thought alike on a particular point.

STRENGTHS

(2) 1. Exposes to a variety of cultures - uncommon in most textbooks.
(2) 2. Stories of students are easily related by the students.
(2) 3. Unique perspective on HoM by bringing other elements of life.

(2) 4. Elements on HoM like the study and research in HoM in India.
(3) 5. Enjoyed reading about a professor passionate about HoM.
(2) 6. The idea of the book is very interesting.
(3) 7. The format and topic of each reflection are engaging, informative and often thought provoking.

8. Each reflection is different. I don't have to read it from the beginning to understand what is going on.

(6) 9. Tons of first-hand experience, personal anecdotes that relate to the topic at hand.

10. Material is relevant to any college student with plenty of advice.

(2) 11. Short and to the point.
(3) 12. Feedback of readers is included in the book - provides unique insight.

13. Sociological, political and linguistic impact of math - interesting thoughts.

(4) 14. Contains information that many do not think about.

15. Succinct - does not drag on.

16. Profound pieces of information.

17. Gives a lot of insight on things you would not find in a normal history book.

(2) 18. Seeing other people's thoughts other than the instructor.

19. Reading your book with your personal insights gave us more knowledge of who you were.

20. Showed a more human/personal side of history.

21. The reflections made me reflect and think about history and the present.

22. The use of flowery language was unique.

23. Since many ideas revolve around UNLV, it feels relatable/applicable to me as a student.

24. He touched unfamiliar topics and exposed students to new ideas.

25. I love reading quick thoughts on a topic that are triggered by some seemingly random incident.

Each reflection is just thought up in the moment and I like seeing how things are connected.

(2) 26. Getting to see into your mind and seeing how you make connections was inspiring to me.

27. Most books are biased and informative and it was really nice to have a book that discussed thought
and ideas.
28. Talks about Math through experience.
29. Complements the class very well.

WEAKNESSES

(5) 1. Repetitive.
(2) 2. Spelling and grammatical errors.
(2) 3. Could use a re-ordering so relevant reflections are in groups.
4. Parts of the section are tangential to the main topic of the section.
(3) 5. Each Reflection should be numbered for quick navigation and search.
6. Slightly seems all over the place.
(2) 7. No unifying theme - many points, but few strings to tie them all together.
(3) 8. Very unsure what I was reading about at times.
9. More information on the topics than needed.
10. Book jumps around a bit on topics and needs a clear goal or better organization.
11. Some chapters could have been left out.
12. Talks more about history than math.

MY BRIEF REMARKS

Some of the positive comments are even beyond my expectations. Yes, at the age of 70, when I started writing non-fiction books, one of the objectives has been to add newness and freshness in respective fields. I intend on beating my own benchmarks.

Since the writing of the very first book, I have taken full ownership of my book. No ghostwriters, no consultants, no proofreaders, no editors, no designers! The main reason being that I have fallen in love with the English language. The only way to master English is to do the proofing and editing of my books – it is a journey into my soul. I am getting better at the language and so are my books! Naturally, some stubborn errors from the typos to spelling still go undetected. You can only catch so many of your errors with your own eyes!

My writing style is my signature. It may not be pretty for some, but that is the way life is. My ultimate focus is on my ideas, which are marinated and brewed for a while. Thus, I am open for a contest of ideas. Furthermore, when a book can be read from any page, then each reflection has to be independent in all respects - thus repetitions are unavoidable. Material has to appear disjoint. In a tweet age, such a format of material is ideal.

May 18, 2017

32. *MATH THROUGH THE AGES - REVIEWED*

In the US colleges, there are two ways of adopting a textbook - one for the multi-section courses and the other for only one section. In the last 40+ years at UNLV, I always go along with the decision of the textbook selection committee, when it comes to any lower division math course - like, college algebra, precalculus and calculus. There being a fierce competition between the publishers, any five nationally popular textbooks are almost isomorphic. As a non-fiction author since 2000, I always wonder at the pricing of the math textbooks, in particular - touching $200 in some courses. It is a national rip-off, as it is in the world of medicines.

Generally, my decision, in choosing a textbook for my course, is rather quick. Mainly, I change a textbook for the sake of overall novelty, and the number and variety of examples and problems. I have no time to go through its details, as my long experience takes over while thumbing through the pages. That is how I adopted ***Math Through The Ages*** (2003) for the **History of Mathematics** (HoM) course (Math 314). The glaring feature that attracted me was its format - a summary of HoM in 60 pages followed by thirty crisp essays, called sketches on common mathematics topics. They are in a broad chronological and curricular order. Most HoM textbooks, running in 700-900 pages, overwhelm the students. However, its 320 pages are inviting to be read and discussed in the class. Also, the familiar names of the Mathematical Association of America as publisher, and of Berlinghoff & Gouvea as authors, made my decision easier.

Overall, the textbook went very well. All along, I used my book***, Darts on History of Mathematics*** (2014) as a supplement. Together, they provided me with the most satisfactory experience in teaching HoM. At the end of the semester, I decided to get the input from the students on each book. They were asked to list at most two strengths and one weakness. In summary, the students liked both books overwhelmingly. The students' detailed review of my book has been posted on Amazon. Below is a summary of their comments on the textbook:

Strengths in snapshots: brief history; fantastic; good primary sources; flexible; can read it from anywhere; story style; short enough to hold your interest and to promote further research; nice learning curve; enjoyable; good problems at the end; history is touched from all over the world; liked history of mathematical symbols; started slow, but got more interesting at the end; great quotes; breakdown of sketches is useful in understanding.

Weak spots and my injections: some students "wanted to see more information on some topics", which is positive - as the textbook had built up their appetite. "Lack of transition between the two sketches", which I think is unavoidable in such a format. "No in depth mathematics," HoM course is not a math course. "Information is interesting, but presented in an unexciting manner"- take it, as it comes.

It is after having used a textbook for a semester and with students' feedback, that objective evaluation of the book is done. Summing up, I will adopt these books whenever I teach this course again. In mathematical jargon, my book has a local character of UNLV, and the other book is relatively 'global'.

May 19, 2017

33. MATHEMATICIAN IN MY BACKYARD!

Years ago, I put my thoughts together as to who is/was a great mathematician. Versions of that write-up are included in the first volumes of my books, ***Mathematical Reflections*** (2010) and ***Darts on History of Mathematics*** (2014). This question came up again during a discussion in the **History of Mathematics** course (MAT 714) that I am teaching this semester. Yesterday, my mental deliberation on this subject took a turnaround of 180 degrees when I read the first half of the following sentence in an email from Brian Winkel: "*We just got word of a $500,000 NSF proposal for SIMIODE project and I have lots of meetings, paper sessions, and mini course activities as well* (also talking of the Joint Mathematics Meetings in January 2017)". This line suddenly stopped me in my tracks of searching the norms of a person being a distinguished mathematician. My feelings were akin to the reaction that one encounters when one goes too far in search of an object, which has been sitting mysteriously under one's nose.

Brian and I have been known to each other since our doctoral days at Indiana University (IU), Bloomington. Brian joined IU two years before I did in 1968. We even took complex analysis courses together. Brian was often seen around the graduate students from India, where he hovered over them with his 6', 5" tall and lanky frame - always enjoying a bellyful of laughs. We got to know each other well enough that in August 1971, I drove my family to visit Brian and his wife, Phyllis in Albion, Michigan. He had just joined Albion College (est.1835), a small 4-year college affiliated with the United Methodist Church.

I do not know how Brian got interested in Cryptology (means secret writing). My guess is that he must have found some applications of group theory, a broad area of his doctoral dissertation. Applications and modeling in mathematics became Brian's passion for life. In addition to teaching and research, in 1977, Brian started a mathematics magazine, ***CRYPTOLOGIA*** - synthesizing teaching innovation, research and public dissemination of ideas on a new subject. Brian told me how his wife and their two kids helped him in every aspect of its production.

Brian was the founding editor and publisher for many years. Once I agreed to review a paper by Subhash Kak that was submitted to the ***CRYPTOLOGIA***.

It was my first review of a paper, which was delayed so long that Brian never asked me to review a paper again! But I did not mind that. My thinking was that since, at UNLV, I was not going to be into the rat-race of publish or perish, so why should I bother about doing any reviewing.

Albion College had a limit on the number of its tenured faculty. Brian was not worried on that account, because he was confident of getting it as and when he would be eligible. However, Brian's friends in mathematics known to him since IU days and working in Rose Polytechnic Institute, Terre Haute, Indiana (est. 1874; now known as Rose–Hulman Institute of Technology), pulled Brian away from Albion after his working there for 10-12 years. Brian flourished there and brought in a lot of grant money to the Institute for the designing of innovative curricula etc. In return, the President gave Brian extraordinary raises at times when the national economy was in a downturn.

Brian told me once that he could thrive professionally in a no-tenure culture, which he advocated for greater faculty creativity. In 1991, he started a second journal ***PRIMUS*** (means first among equals). The history of most mathematics journals is that they are sponsored by the big-time publishers, who get well-known mathematics professors as chief editors in order to draw journal subscriptions. Brian's story is unique in the sense that an enterprising young math professor had founded two successful journals while working in no-name schools too. In 2007, the publishers, **Taylor and Francis** purchased both the journals. Brian was happy with the financial deal.

The nature of mathematics is absolutely self-centric. Most math faculty spend their entire lives in a single dimension - teaching the same courses, serving on the same committees and doing incestuous research for a while from their dissertation days onwards. It seems Brian has been reinventing himself intellectually every 10-12 years. The story of his quantum leap from an engineering institute to the United States Military Academy, West Point (est. 1802) is very fascinating. It is not easy for a 50-year old civilian to start working with professionals in uniform and take directions from them. However, Brian blossomed there too. It speaks of his ability to work with people with different temperaments, ethics and cultures. I know a person who left West Point disgruntled after three years.

After selling off both the journals in 2007, Brian transitioned into a digital domain. In 2013, he founded SIMIODE, a 501(c)3 non-profit organization based in Cornwall, New York. SIMIODE is supposedly a rich environment for learning and teaching differential equations through modeling. SIMIODE, the longest acronym that I have run into, stands for the Systemic Initiative for Modeling Investigations and Opportunities with Differential Equations.

SIMIODE is truly a baby of Phyllis and Brian conceived after Brian's retirement from West Point. Phyllis has been on Brian's side ever since she took a liking for him when he was a clumsy college boy working as a lifeguard. Two years ago, for a surprise celebration of their 50th wedding anniversary, their son, a business professor, approached me to write a reflection on this occasion. What a pleasure it was!

For the last few years, I have been seeing Brian's following signatures at the end of his emails:
Brian Winkel, Director - SIMIODE - www.simiode.org
Editor Emeritus - PRIMUS Editor Emeritus - CRYPTOLOGIA
Professor Emeritus Mathematical Sciences, United States Military Academy, West Point, NY, USA.

Strangely, it was only yesterday that these signatures revealed their magic on me. Brian may not have served mathematics by doing traditional research, but he has already made significant contributions in mathematical applications, teaching, commerce, and curricular innovations through **CRYPTOLOGIA, PRIMUS** and **SIMIODE**. To the best of my knowledge, no one matches this mathematical record.

I am so glad, surprised and proud to see Brian perched up at the top of his own mathematical mountain! On a side note, Brian is the only mathematician who is Emeritus Cubed! This is what the American dream is all about - to be your own boss.

October 30, 2017

34. WRITING TEXTBOOKS & PUBLISHING PAPERS

In a mathematical jargon, what follows is a well-posed question: Do the set of math faculty writing collegiate level textbooks and the set of math faculty publishing traditional research papers have a non-empty intersection? This English sentence is correct and clear for all intents and purposes, but its answer is not unique, as several x-factors are involved in this question.

The most important x factor is the caliber of a math department and ranking of a university. For instance, the expectations in mathematical research, both qualitatively and quantitatively, are very different at universities like, Harvard, Princeton and Stanford than they would be from the corresponding expectations in universities, say, like, UNLV - my professional home since 1974. Also, for ages, mathematicians have been arguing over whether research in Area A of mathematics is more challenging than research in Area M, and so on.

Generally, the writing of textbooks at the undergraduate level is different from the ones at the graduate level. On the other hand, publishing a research monograph is a kind of double dipping, as often, it includes 80-90 % of the author's research papers, which are already published and/or presented. Nevertheless, some merit of its consolidation is not to be denied. But, the writing of a math textbook at the lower division level is an institutionalized plagiarism, as it has mathematical creativity of order epsilon.

On a personal note, for the aforesaid reasons, I vowed never to write a mathematics textbook. As far as publication of research papers is concerned, I took myself out of the Publish-or-Perish rat race, as soon as I finished my PhD. I was then 34 years old, married - wife, and children - eight and five-years old. My married life had cracked over in my pursuits of a PhD in India and the US - for 12 years since 1962. Ever since I read GH Hardy's classic book, ***A Mathematician's Apology*** (1940), I was convinced that my golden years of research were way past. Therefore, instead of spinning the wheels of mediocre research, I decided to find my own niche in multi-dimensional scholarship. Driving through the vast lands of the US - from Indiana University to UNLV, turned out to be an exploration of the West - both literally and figuratively.

Well, this train of thought hit my mind when I was reading this morning the Foreword, two Prefaces of two editions, and a write-up, **What is Mathematics?** of the book incidentally by the same title, ***What is Mathematics?*** (1941). Its subtitle is - ***An Elementary Approach to Ideas and Methods***. The value of a book is measured by new insights its reading gives to a reader after every few years. In that respect, all religious scriptures are unique, as they are read, interpreted, and commented upon so frequently by their respective adherents, and scholars at large in every land and age.

The ideas in ***What is Mathematics?*** mainly belong to Richard Courant (1888 - 1972); however, his collaboration with Herbert Robbins (1915-2001) was significant in its finish. Having moved from Germany to the US in 1936, Courant's command over English had not developed enough for the writing of such a book in English. Its current and second edition (1996) was also updated by Ian Stewart (1945-), a UK mathematician. It is not out of place to add that besides these two top notch mathematicians who involved themselves in this book, it was reviewed by Albert Einstein (1879-1955) and Herman Weyl (1885- 1955). This sets the book apart from others - seven decades validate its intrinsic merit.

Let me clarify that ***What is Mathematics?***, being used for the fourth time, is a unique 'textbook' for the 50% of the material covered in MAT 711-712 (**Survey of Mathematical Problems I and II**). This sequence of graduate courses is exclusively designed for **Teaching of Mathematics** concentration, which is one of the four standard concentrations in graduate programs in mathematics.

Talking of ***What is Mathematics?***, Courant wrote in the Preface of the First Edition that **"*The book is written for beginners and scholars, for students and teachers, for philosophers and engineers, for classrooms and libraries*"**. It is so beautiful, and reads like a line of poetry. The book is absolutely not a run-of-the mill mathematics book. Courant has also written two volumes each on ***Introduction to Calculus and Analysis*** and a classic, ***Methods of Mathematical Physics***. This is quite a body of mathematics books to be reckoned with.

The next question is with regard to what kind of research mathematician Richard Courant was. The answer is very simple, as his name still Googles up with three key results in Applied Mathematics after his name; a passport for immortality - namely, Courant Number, Courant–Friedrichs–Lewy Condition and Courant Minimax Principle. My retrospective conjecture is that Courant would have won a Fields Medal and an Abel Prize, if they had been instituted when he was alive and he had been eligible for them.

Apart from being a great mathematician and a significant writer of mathematics books, Courant was such an able administrator that the Mathematics Department of New York University that he joined in 1936, was named after him in 1964, as the Courant Institute of Mathematical Sciences. Such a recognition to a living individual happens very rarely.

In conclusion, the answer to the question posed in the beginning of this reflection is that the two sets are technically not disjoint. I would also include PR Halmos (1916 - 2006) and John B. Conway (1939 -) in the intersection of these two sets. If one sees some bias in my favoring Conway and Halmos, then frankly it is because I took courses from them when they were on the faculty of Indiana University, Bloomington. However, the intersection of these two sets may not be statistically significant. Or, in the jargon of Analysis, the intersection of these two sets is empty modulo a set of measure zero!

Viva ***What is Mathematics?***

August 29, 2018

COMMENTS

Satish. Really very nice essay about the timelessness of good books. Keep up the good work and stay cool now that school has started. **Brian**

Thank you for sharing your reflection. It is interesting to read. I agree with what you said about the low level books. As for research monographs, yes, they mostly repeat previously published papers, but I do not see it as a bad thing. Usually these books will be written by a person or a group of researchers who believe that they have found a new approach or technique, built on it, applied it in their research and now want to share it with others. If it is a comprehensive, thoughtful work, such a monograph can serve as a good resource for graduate students and also for the people who have recently started working in this particular research area. So, the interception is not empty? More than that, the best books like Courant's book are usually written by people well known for their research. The other thing, when they are active in research, they probably are not that interested in book writing. Does it mean a disjoint or a joint set again? Thank you one more time for sharing the reflection, and have a good weekend! Best regards, **Viktoria**

Dear Dr Satish, I am very thankful for such an informative email and I quite enjoyed reading it. **Sabahat Malik**

35. CELEBRATION - INSPIRATION!

In the fast and competitive world that we are living in, there is little time to pause and celebrate the historic milestones and achievements of a colleague or of a co-worker. Digitization has added a layer of alienation to life. However, it is a sign of a mature institution that celebrates its living members while growing up in size. Human values define a university too. Well, these thoughts welled up in my mind when I learnt that my colleague, Sadanand Verma turned ninety (90) on January 24 - yes, three days ago. He is the oldest full time faculty member at UNLV. A perspective on his life span is that UNLV, founded in 1957, is only 63 years old.

Verma has scored many firsts in his illustrious and still ongoing career at UNLV (was called Nevada Southern University until 1969). He joined the Mathematics Department as a (full) professor in 1967, and was the first Math PhD holder. At that time, there were hardly a dozen math faculty. Most of them had master's and two had doctorates in math education. My guess is that unknowingly Verma may have set a US record for being the oldest and yet, active tenured full professor of mathematics. Apart from longevity related milestones, Verma's contributions to UNLV are very significant. Here is a partial list:

1 Verma served continuously as the Department Chairman for 22 years (1968-1990) - a record at that time. Generally, the chair's term was two years until 1997.

2. Soon after Verma's joining UNLV, the master's degree in mathematics was initiated and quickly approved in 1968. A sidebar: UNLV's present Graduate College, known then as the Division of Graduate Studies, was established in 1964.

3. Verma was visionary as he saw computer science galloping on the academic horizon. He just started offering computer programming courses. As their demands increased, he got new faculty positions. During 1974-80, I was one of the seven PhD faculty who were hired. I did teach some computer science courses too!

4. Under Verma's leadership, students started getting bachelor's and master's in computer science as concentrations. With growth comes new administrative alignments. The Computer Science 'group' joined the Engineering group when the **Division of Engineering** was carved out of the then **College of Science, Mathematics and Engineering**, which was naturally renamed as the **College of Science and Mathematics**. Giving a touch of institutional history, in 1998, as a part of the campus reorganization, Mathematics was dropped from the College's name.

5. The history of lecturers' positions has been convoluted at UNLV. In 1985, Verma convinced the then Provost, John Unrue, and got the first four lecturer positions (non-PhDs) for the Department for teaching remedial and lower division courses. Gradually, the number of lecturers doubled up by 1990.

As I turned eighty (80) last month, I am delighted to share this reflective piece with the community at large so that they may know that Sadanand Verma is a pioneer and builder of UNLV right from the grassroots level.

January 27, 2020

COMMENTS

Satish, Wonderful reflection on a dear and cherished colleague and on the contribution to the process of universities folks of all ages can make. Thanks for sharing. **Brian**

PS: I am ever so glad to have our friendship and I can always hear your voice and see your smile, saying, "Oh Brian, I tell you . . ." and then something interesting happens. Take care and regards to your bride and family!

Very nice. You made my night. **Joe McDonald**

Thank you Satish- This is beautiful. I appreciate you sending it. Happy happy 90th birthday to Dr. Verma AND happy happy 80th birthday to you. All the best, **Michelle Robinette**

Hi Satish, Thank you for sharing this well written and well deserved write up for Verma. He is amazing. **Angelo Yfantis**

In his latest blog posting, Satish just revealed it is Dr. Verma's 90th birthday! So, the happiest birthday wishes from me. I was going to speak with him privately during my holiday visit but, of course, I didn't make it last year, so I'll do it publicly. I just wanted to tell him how much he has meant to me over the past five decades: first by getting me a job as a paper grader while I was an undergraduate, then hiring me as a TA (the first one?) in graduate school, hiring me as a part-timer when I got out of the Army, hiring me as a Lecturer/Instructor for ~20 years (I call it 20 years.), and guiding me through many years as our Chair. Thank you more than I can ever say. And I apologize for not mentioning this earlier. **Hal Whipple**

36. MATHEMATICAL FOUNDATIONS OF MY CREATIVE WRITINGS

[**Note**: This reflective article is modified from a paper that I presented during the 32nd annual conference of the **Far West Pop Culture Association** held in Las Vegas during February 21 - 23, 2020.]

Scope of my writings

In 2010, at the age of 70, when public perception of most professors is of deadwood, I published my very first book - which was mathematical in nature. It has many angles at mathematics, but it is miles away from being a mathematics textbook. As a matter of fact, years ago, I had ruled out the writing of any lower division undergraduate mathematics textbooks as in my mind that is institutionalized plagiarism.

In 2012, my banner year in writing, there came out three books in three different genres - namely, historical, philosophical and religious. Subsequently, a couple of books were classified in literary and educational genres. Thus, by 2017, I had rolled out ten books in six different genres. All of them are unique in content as well as in format. That is a hallmark of original work by any standards. However, I am not yet done writing with books, as at least five books are in the pipeline - God willing.

Being 81 years old now, it no longer behooves me to talk about somebody else's creative work. After all, what does a long life span mean, if a claim to some creative work is not made! Before expanding on my claim, it is pertinent to paraphrase on creativity in a broad manner and set it on a common ground with general readers. The following 'parameters' are set in a caption layout. This brings greater clarity on the subject and may point out to the soft spots in it. Let me start with the following obvious question:

What is Creativity?

One can go on forever in defining, arguing and exploring creativity. Briefly speaking, it may not be definable in every aspect of life, but often it is not difficult to be recognized. In my world of mathematics, there is a consensus that seminal work is done before the age of 40. Mathematics is like boxing, where a world championship is seldom won after the age of forty. For that very reason, the Fields Medal, the highest honor in mathematics, awarded every four years, goes to mathematicians under the age of forty.

Creativity in Human Beings

As biological entities, human beings are unique. Consequently, each person has the potential for doing some original work. At the other extreme on a social totem pole, serial killers, IT hackers, stock market manipulators, odds fixers, 9/11 terrorists, and abductors for ransom are creative too, but in unlawful and anti-social territories. Also, not discounting the creativity of the gymnasts and athletes who perform at a super human level that inspires as well as entertains people.

Genesis of my creative writings

My present writing is a reincarnation of the old-fashioned letter writing that goes back to my teen years. Unknowingly, I may have set a world record by having written the longest letters (a couple in 100 pages), saving thousands of letters written to me, and having read letters of most men and women of name and fame. By and large, a person is truthful in his/her letters that are written to his/her near and dear ones. However, once emails came upon the communication scene, letter writing met its natural death. In Y-2000, out of a writing vacuum of the 1990s, there burst open my dormant fountain of reflective writing. My one-to-one mode of letter writing became one-to-many in reflective writing; in a way from local to global, and from intimate to impersonal. However, what has not changed is my passion for writing.

Choice of Mathematics

In 1961, I made a conscious decision in choosing my lifelong profession of mathematics teaching at the college level. Since then, mathematics has permeated into every pore of my intellect. Mathematics solves all kinds of problems. Consequently, all my ten books are in the non-fiction domain. Here comes a quantum shift of thoughts. Once, one has been in a profession of one's liking for 30-40 years, then the characteristics of that profession start rubbing off in one's persona! Thus, my mindset has been significantly molded by mathematics.

Mathematical Foundations of my Books

Mathematics is a formidable, though, a man-made discipline; a powerful tool, as real as a universal wrench in a garage. It has been known as a language of sciences since the time of Galileo, but now it is of all disciplines! Also, there is a 'nuclear power' of mathematics that lies in its deductive reasoning - the essence of mathematical thinking. However, this power of deductive thinking in diverse areas can only be unleashed by mathematicians who have unshackled minds. It is not easy to decondition the binary nature of a mathematical mind, and then divert its energy into multi-disciplinary channels. It requires concerted effort over a long period of time. What follows is a gist of my books in each genre.

1. Mathematics and Connections

Mathematics being pervasive in life, my two volumes on mathematical reflections connect mathematics with topics traditionally remote from it - like, boxing, arts, dance, music, communication, football. Themes include like, Calculus Defines Civilization, Philosophy turning Mathematical, Birthing Pains of Proof, Teaching Awakens Creativity, Holocaust and Godel's Theorem, Skewness of Linear Algebra, A Convergence of Slavery and Mathematics, Broccolis, Carrots & Math, A 4-Dimension Calisthenic, Mathematics of Fun, Culture Affects Creativity, Prayers and Research, Relativity-Science-Divinity. In popular jargon, Mathematics is like a world conqueror, an imperialist, and a great colonizer - making connections with any human activity.

2. Mathematics and History

Mathematics being grounded and objective, the facts in my history books are more 'factual' due to the deductive nature of mathematics. History is also about interpretation of events - connecting various dots. In mathematics, any two points can be joined by infinitely many paths. However, it requires deconditioning of the mathematical mind before dwelling and delving into the writing of history books. The two volumes of **Vectors in History** have been a challenging undertaking.

3. Mathematics and Philosophy

The edges of every discipline lend to philosophical speculation. Since 1995, I have been a chartered member of the **Philosophy of Mathematics**, a subgroup of the **Mathematical Association of America**. Philosophy deals with abstraction and so does mathematics in all its concepts. At one extreme, in set theory, mathematicians deal with not one infinity but with many kinds of infinities! In such a study, one gets out of breath and gasps for oxygen. Since the eras of scholars of ancient Greece and of ***rishis*** of ancient India, mathematics and philosophy have been in bed together. The first volume of, ***Epsilons and Deltas of Life: Everyday Stories*** deals with Applied Philosophy.

4. Mathematics and Religion

Rational pursuits of infinity in mathematics is nothing less than an understanding of the Divine, divinity, deities or manifestations of God. Mathematics alone lets one have a cogent glimpse of the Supreme, whatever image one has of it. **My Hindu Faith & Periscope** was the most challenging book for me to write. It amounts to standing on the solid grounds of mathematics while peering through the issues of life from the lens of my Hindu faith.

5. Mathematics and Literature

Mathematics is a language and has its own grammar - the rules of so-called manipulations are very strict. There being no ambiguities in the definitions and notions of mathematics, my non-fiction literary writing tends to be concise and compact. At times, a paragraph in my writing can be easily expanded into 2-3 paras. Apart from the structure of writings, there is a realism in it, which oozes out of mathematics.

6. Mathematics and Education

Basically, my book, ***Plums, Peaches and Pears of Education*** is an outgrowth of my being in the education enterprise for nearly six decades. Mathematics provides incisive thinking at various levels of education. The domain of education is vast and varied. However, math education, my current area of research and scholarship, takes care of a good segment of education. This book is largely drawn from my experience of frequent traveling and teaching in various countries and educational systems.

In conclusion, to the best of my knowledge, Bertrand Russell (1872-1970) is the only world-class mathematician, who has been a multi-dimensional mind in recent times - a polymath, a renaissance person. He has been my mentor-at-large since my high school days. Russell was the most sought-out speaker in the world, a social iconoclast, a Nobel Laureate in Literature (1950), global pacifist, and a nominee for the Nobel Peace Prize. Russell would have won the Fields Medal in mathematics too, but this award was instituted in 1936 when Russell was 64 years old - way past the age limit of 40! Obviously, the converse of the above six 'theorems' is not true!

February 29, 2020/March, 2021

37. REUBEN HERSH PERISCOPE

It is a good editorial policy to publish eulogies and profiles of mathematicians no matter where they are from. At the same time, it is essential, if some features of their lives can be connected with present academic conditions in India. Here are my brief comments just after reading a synopsis of the rich life of Reuben Hersh (1927-2020) as it appeared in the July newsletter of the ***Gujarat Ganit Mandal*** (Gujarat Mathematics Circle).

1. How can Indian colleges and universities have a flexible curriculum in which a person, say with a degree in English or Hindi can easily also go all the way to do PhD in math and become a solid mathematician? Hersh was also able to combine his assets in English and Math to write a few popular math books and many articles for magazines. In 1983, I used his first book, ***The Mathematical Experience*** in teaching a course on **Humane Mathematics** (Math101) at UNLV.

2. Even after 70+ years of India's independence, education at all levels remains fragmented - too much and too early specialization. Consequently, the professionals generally are burnt out before blossoming up. Professionally, they are bored and end up living in dismal mediocrity for a good part of their lives. For the Fields Medal or the Nobel Prize type of research work, Indians always look outside India. This is a state of mathematical creativity in India.

3. Indian mega corporations like Reliance, Tata and Modi can make a small dent here. In each round of hires, they should experiment with hiring a few persons based upon their street experience, self-study, self-readiness - measured by the 10000-hour rule (Malcolm Gladwell).

4. In India, there is still a misconception that if a program of study of 2, 3, or 4 years takes a year or two longer, then additional time is considered to have been 'wasted'. Again, looking at Hersh's life, as a young adult, he worked as a machinist on his way to mathematics.

5. Being a charter member of the **Philosophy of Mathematics**, a subsection of the Mathematical Association of America, started off in 2000, I often saw

both Reuben Hersh and Philp J. Davis (1923 - 2018) at the annual meetings. It may sound amusing to write that I have generalized Hersh's thesis of **Humanistic Mathematics** by bringing all disciplines of humanities and social studies in the orbit of mathematics. That is the approach I take in teaching courses on History of Mathematics.

6. It is ironic that History Mathematics is still nowhere in the academic horizon of India. **History making, history writing, and history reading measure the power of a society or of a nation**. No wonder universities in the UK and US lead the world in the study of History of Mathematics. It is high time to push for a semester of history of mathematics in colleges and universities of India. Private colleges and universities can take this lead. A saying goes - 'History is for the wise'.

Thanks, Prof Vasavada (Editor of the newsletter of the ***Gujarat Ganit Mandal)*** for your selfless service to the cause of Mathematics for over four decades.

July 01, 2020

38. ARUN VAIDYA - 'FIRST' GUJARATI MATH HISTORIAN!

I am not a prophet, but I am prophetic at times. Prophecies could be similar to statistical inferences too. At the age of 80 Plus, one does accumulate a lot of data in every aspect of life which gets stored in the deep vaults of one's brain. With a sound memory retrieval system, one can foresee the shadows of some coming events. The flashing of Arun M. Vaidya/AV (1935 - 2020) in my mind during the last two days is one of those premonitions that has materialized today. However, for all intents and purposes, the universe is mathematical to me.

Here is a sequence of my clairvoyance on AV. Last Monday, while compiling the material for the second volume of my forthcoming book, ***Darts on History of Mathematics*** (HoM), I decided to do a minor reorganization. The first four sections of the book have nearly 70 varied reflections on HoM - all written by me. However, the fifth section includes perspectives on HoM by mathematicians whom I personally know that they have developed some sense of history too. I did not want to include such views as published in textbooks and articles. My plan is to reach out to 2-3 mathematicians from the US and 2-3 from India. That is how AV came up on the Indian list. The irony of life is that instead of asking him to submit an article in at least 300 words, I am writing this eulogical reflection on him!

In trying to remember the day when AV and I first met or when we communicated with each other, that is not clear to me at this moment. A sidebar: I had seen his uncle, Prahalad C. Vaidya (1918 - 2010) a couple of times during the annual meetings of the Indian Mathematical Society. For years, I thought AM Vaidya was the son of PC Vaidya - knowing fully well that of all the biological traits, the chances of math genes passing from father to son are very slim.

However, AV has been on my two mailing lists of reflections for the last 10-12 years. Apart from exchanging emails, we chatted on the phone once or twice a year. Once I started working on another book, **History of Mathematics of Punjab**, I urged AV to undertake the writing of a similar book on Gujarat. Instead, AV took upon himself the writing of a book on the life of his uncle, PC Vaidya. It was by God's grace that he survived a massive stroke in 2015 that had kept him in intensive care for a month. Eventually, he fully recovered, and he was able to finish this biography, and see to its publication.

Well, AV did not get on to the project of the **History of Mathematics of Gujarat**, as he felt intellectually drained after the long standing biography was published. He specifically admired my pieces on HoM in Punjab. The purpose of regional histories of mathematics is to kindle interest in history in general in which the Hindus have stayed miles away. Without the knowledge of history, one is out of politics too.

In a non-mathematical vein, when AV and I met ten years ago in a HoM conference held in Vidyanagar, AV gave me rides to and fro from Ahmedabad. I told him that his fair complexion and sharp features were due to the bloodline of the Hindus who exodused Kashmir during their persecution by the Muslim rulers from Afghanistan, Iran and Central Asia. They took refuge in the Hindu kingdoms in south India then - comprising the present states of Andhra Pradesh, Gujarat, Karnataka, Kerala and Tamilnadu. AV just smiled. Long live the legacy of Arun Vaidya!

November 26, 2020

39. DARK SPOTS ON THE MOON (PART III)

[Note: This piece is extracted from a 21-page compilation of short essays - my ongoing project on the history of mathematics. I have essentially pooled 200-300 word submissions of the students in several math courses. The students earn extra credits for doing it. It is not mandatory, but most students do it. For the purpose of illustration and space, ten have been chosen for this volume. For a historical perspective, they are put in a chronological order. However, I have not changed much of the students' language. Sometimes, one is surprised as to how a new cooking recipe would turn out; a line of research shed light on a different problem; a hike opens out to exhilarating new vistas. A month ago, an article in the ***Focus,*** a newsletter of the MAA prompted me to examine the lives of legendary mathematicians from an 'opposite' end. As I enter into my 70s, it is becoming increasingly clear that every life approaches its zero sum in terms of its highs and lows. Their research has stood the test of time, so there is no commentary onit. The focus is on their frailties and failures in other walks of life. The students were then inspired by their achievements despite other problems and hurdles.

1. Johannes Kepler (1571 – 1630)

Johannes Kepler was a mathematician and astronomer. His father was a mercenary and left the family when Kepler was five years of age. His mother earned a living as a healer. His primary education was done in a protestant environment; with the belief that ultimate authority and say is given to the Bible rather than to/by any man. Many claim that Kepler's upbringing in protestant faith played a huge role in his way of thought and motivations to derive many of his theories. After completing his education, Kepler became a mathematics teacher and later got involved with a variety of occupations including the assistant astronomer to a senior well known astronomer Tycho Brahe, and eventually the imperial mathematician to Emperor Rudolph II.

Contrary to what one would believe, he did not primarily study math or astronomy in the university but rather studied theology and philosophy. Amazingly, the reason he succeeded in becoming a great mathematician. Kepler had many other milestone theories but all of them had one origin, one motivation. Johannes Kepler was a firm believer that God created a world according to an intelligible plan that is accessible through the "natural light of reason."

2. Blaise Pascal (1623 – 1662)
Pascal was a man of many mysteries and intriguing life. He was more than the Pascal Triangle. Pascal's teacher was his dad, who taught him everything; however he refused to teach him mathematics. This refusal led to Blaise's curiosity to mathematics, which eventually led to his greatness in mathematics.

His very first invention was a mechanical calculator, a gift to his father. Due to poor health he became hypochondria, a phobia of excessive worry about having serious illness. Immediately, after his father's death, his sister left Blaise on his own. Blaise begged for her to stay. As a result, he developed a fear of abandonment. His last words prior to his death were "May God Never Abandon Me." According to his other sister Gilberte, Blaise was disgusted by talks on feminine beauty.

After an accident, he was involved in; a religious vision that came forth. What was fascinating about this vision was that he had sewn the note on his clothes. Every time he changed clothes, he would sew it back on a new shirt. His servant had only noticed this strange behavior after he died. Descartes was not convinced that this young boy could do precocious works; consequently, they eventually became rivals.

3. Isaac Newton (1642 – 1727)
The man known as Isaac Newton was a wonderfully gifted and equally a mysterious man. Throughout life, he suffered from social problems. One of the biggest themes about Newton was that he actually found that the things that most people thought make life worth living were nothing more than distractions, especially, women with respect to romance. It would almost seem that he wanted to keep his research for himself, but also that he did not want others to take credit for the things he figured out first. One last thing that a lot of people didn't know about Newton was that he practiced Alchemy, possibly the most controversial thing in Newton's life.

Newton's father had died before he was born. Three years later, his mother remarried and moved into the home of her new husband, Barnabas Smith.

Being at the age of three, infant Newton was left to the care of his mother's parents. About seven years later, Smith died, and Newton's mother came back with her two daughters and a son from her second marriage. This childhood experience would permanently damage Newton's ability to trust others. He would go to the tavern and occasionally play cards. In one of these manuscripts, Newton listed all of the sins he had committed, and it was addressed directly to God. Some of the 'sins' were 'Striking many', 'Punching my sister', and 'Wishing death and hoping it to some'. These sins represent early signs of future rages and dark depressions that will follow Isaac Newton's throughout his adult life. Although Newton had become rich, and overloaded with honors, he remained deeply insecure, given to fits of depression and outbursts of violent temper to those he felt threatened by.

4. Gottfried Leibniz (1646 – 1716)

Although Newton's popularity and political pull now appear to be the primary forces behind the accusations, there are, in fact, a number of suspicious details casting suspicion on Leibniz's claim. Additionally, Leibniz is known to have deliberately altered important documents in other instances. He has admitted to altering his April 8, 1716 letter published in ***Acta Eruditorum***, and the letter of June 7, 1713 in the ***Charta Volans***, and to changing the date on another manuscript. His possession of Newton's manuscript can perhaps be dismissed as evidence, but this history of deliberately altering published documents is very compelling character evidence. Before Leibniz's death in 1716, a Royal Society committee officially ruled in Newton's favor on this matter.

5. George Peacock (1791 – 1858)

George Peacock is also known for his efforts to reform the teaching methods at Cambridge University, where he studied. One fact of note is that, as a child, he was not very good at math. It is often common to think that someone who grows up to become a world-famous mathematician was some sort of child prodigy in mathematics. He was quite an athletic child and was more remarkable for his ability to climb high distances than for his intellect. Peacock is a perfect example that one does not need to have excellent math grades in school to do great things in the field. Mathematicians often get a reputation of being nerdy or un-athletic. An interesting fact about Peacock is that after he died in 1858, his wife Frances Elizabeth married one of his students, William Hepworth Thompson.

Peacock's devotion to his church was so deep in his life that he served as a deacon, priest, Vicar and Dean.

6. Emmy Noether (1882 -1935)
Emmy Noether was the daughter of a well-known Jewish Professor in Germany. Her parents had high standards for their children. Emmy's successes in life can be measured by her parent's ambitions. Emmy also worked at the University of Gottingen without pay or position for 8 years and published articles anonymously through a friend. Six of them are now considered classics. In 1933, Germany's Nazi government dismissed Jews from university positions, and Noether moved to the United States to take up a position at Bryn Mawr College in Pennsylvania. In 1935, she underwent surgery for an ovarian cyst, despite signs of a recovery, she died four days later.

7. Srinivas Ramanujan (1887 -1920)
Ramanujan's religious fervor, the domineering influence of his mother, and his stubborn nature were among the main influences in his life. In his early childhood days at Madras, he did not like the schools and would find ways and means to avoid going to school – to the extent that his family had to enlist a local constable to make sure he attended school.

Married at the age of 9, but consummated at 17 or 18. Ramanujan had developed a hydrocele testis, an abnormal swelling of the tunica vaginalis, an internal membrane in the testicle. The condition could have been treated with a routine surgery, but his family did not have the money. Finally, a doctor volunteered to do the surgery for free. He spent nine years at home, without a job or a degree, having failed his BA due to his unwillingness to study what he didn't want to. No key decisions in his life were taken without God's blessings. Ramanujan credited his acumen to his family Goddess, Namagiri, and he looked to her for inspiration in his work.

Paul Erdős has passed on GH Hardy's ratings of mathematicians. Suppose that we rate mathematicians on the basis of pure talent on a scale from 0 to 100, Hardy gave himself a score of 25, J.E. Littlewood 30, David Hilbert 80 and Ramanujan 100.' Further Prof. G.H. Hardy of Trinity College, Cambridge went on to claim that his greatest contribution to mathematics was '**Discovering Ramanujan**.'!

8. Kurt Gödel (1906 - 1978)

Nicknamed "Mr. Why" by his family, Kurt Gödel was always an inquisitive person even as a child. Gödel excelled in all subjects in school. He originally wanted to study theoretical physics but added math and philosophy classes. By the time he attended Vienna, he had already mastered university level mathematics. In June 1936, Moritz Schleck, the man whose seminar possibly pushed Gödel into mathematics, was assassinated by a pro-Nazi student. The death of Moritz Shleck affected Gödel in a very bad way. He soon became very paranoid and fell into a deep depression. Because of his paranoiac symptoms he spent several months in a sanatorium for nervous diseases. In spite of his depression, Gödel still carried on a regular life. He got married to a dancer named Adele.

Because of the Nazi regime he was unable to continue teaching and lecturing at the University of Vienna. He decided to move to the U.S. to continue teaching at Princeton University. While there, his paranoia grew but he was still able to formulate theorems. Near the end of Gödel's life he had become so paranoid that someone was trying to poison him that he wouldn't eat any food unless his wife would taste it first. When his wife Adele became hospitalized, Gödel refused to eat because of his paranoia. Gödel eventually died of malnutrition and weighed only 65 pounds at the time of his death.

9. John Nash (1928 -2015)

John Nash won the 1994 Nobel Prize in Economic Sciences. However, Nash had hidden mental and psychological problems that were masked. The first of these controversies began in the mid -1950s. He was arrested in a bathroom in Santa Monica on moral charges related to homosexuality. This caused him to lose his position with the RAND Corporation. In 1957, he married Alicia López-Harrison, a student from El Salvador whom he met as an instructor at MIT in 1951. John 'believed' that men in "red ties" were chasing him and even wrote letters to foreign embassies saying that he was forming a world government. In 1959, while pregnant, his wife had him involuntarily admitted to a mental hospital for paranoid schizophrenia and depression. These complicated their marital problems that eventually led to Nash and Lopez-Harrison's divorce in 1963, stayed in touch with each other, and they remarried in 2003.

It was not until 1995 that he felt that he was finally thinking rationally as a scientist should; although he finally felt sane, he deemed that his genius was being limited. Ironically, both John and Alicia died in 2015 in an auto accident involving a taxi that was bringing them home from the Newark airport after receiving the Abel Prize (Applied Mathematics) in Norway.

10. Grigori Perelman (1966 – alive as of 11/2020)
As a child, Grigori was encouraged to read, write and challenge his mind in math problems. In 2006, he was awarded the Fields Medal which is the math version of the Nobel Prize. He declined to accept it - including a purse prize of $7,000. Earlier, he refused to accept the prestigious prize from the European Mathematical Society. Likewise, he declined the Millennium Award and one million $$ that goes with it. Perelman quit his job and dropped out of the math world completely. He now lives with his mother in complete poverty. Friends of Perelman say he is a simple guy "that a first glance you would think he is a bum because he wears the same clothes every day, lets his hair grow wild and doesn't cut his fingernails"

SECTION III

INDIA SPICES

40. MATHEMATICS IN THE VEDAS & YOGA

[**Note November 2020**: This paper was presented at the First World Yoga Conference held in New Delhi (India), during December 29-30, 1986. It was a phase of my life when publishing a book was nowhere on the horizon. Most of the time, articles and essays written for various occasions and meetings were not saved in any systematic manner. During the last ten years when the book publication became my sole medium of dissemination, I started digging old folders in different browsers and discs etc. Thus, any nugget, surfacing up and still deemed fit, is included in this book. However, I have maintained the basic format and text of the paper.]

All over the civilized world, the Vedas (four in number) are acknowledged as the most ancient Indian scriptures full of wisdom and eternal truths. The Hindus, particularly of the Arya Samaj leaning, go a step further. They believe that the Vedas are the revealed works of the divine forces. Hence, they are considered to be the storehouse of all knowledge necessary for mankind.

Mathematics is not merely a tool for the investigation of physical sciences, but has characteristics of a language too. A solution of any problem when handled with mathematics provides precision, authenticity and newer insights. Shankaracharya Bharath Krishen Tirath (1884-1960) is the first scholar in recent times who took the giant step in demonstrating that true realization is actual visualization, and hence it has to be mathematical too!

In his famous work, ***Vedic Mathematics***, Shankaracharya Bharath Krishen Tirath [1] has put down sixteen ***sutras*** (aphorisms) which he claims to have discovered from the Appendix of the **Atharva Veda**. However, they have not been deciphered in the present recensions of the **Atharva Veda**. Neither Shankaracharya has given any specific reference to some Vedic ***mantras***, nor has anybody else traced their direct origin. The ***sutras*** in themselves are fantastic in their mathematical applications. The questions which arise in my mind are the following:

1. Without any reference to a specific source in the **Atharva Veda**, how did Shankaracharya hit upon those ***sutras***? Was he able to figure out the multiplicity of meaning of the Vedic mantras?

2. Can a person with modern education ever construct a ***mantra*** with multiple meanings? For example,

गोपीभाग्यमधुव्रात - शृङ्गिशोदधिसन्धिग ॥
खलजीवितखाताव गलहालारसंधर ॥

has three distinct meanings braided in it. The first meaning is a hymn to Lord Krishna, the second one is a similar hymn in praise of Lord Shankar, and the third one is the valuation of π (pi) to 32 decimal places with a built-in key to any place! $\pi/10 = .31415926535897932384626433832792$----. According to the History of Pi [2], its approximate value up to 32 decimal places was first computed in 1596 by Ludolph Van Rooman. Only after the invention of calculus and the advent of computing machines, and electronic computers after World War II such approximate values of pi could be easily computed.

When one searches for the answers to the above questions, then one goes into the realm of Yoga - that stands for the union of a soul with Supersoul. It is thus the awakening of all latent physical and spiritual forces in order to expand the horizon of man's holistic understanding. Time and again, scholars - including Shankaracharya himself, have attributed this discovery to his eight years of ***tapa and sadhna*** (deep and long meditation) in the forest of Sringeri. Besides, he was a rare scholar of mathematics and Sanskrit. It was through intense meditation and concentration of mind that Shankaracharya distilled out these sixteen ***sutras.*** He referred to True Realization by means of Actual Visualization. Actual Visualization is the total enlightenment. In the domain of spirituality, this is the awakening of ***Kundalini*** (energy center) or the opening of the Third Eye.

I witnessed another such encounter in a meeting with Pandit Kashi Ram [3], a well-known scholar of astronomy and astrology based on Vedic

Mathematics. He claims to have discovered similar results after years of concentration and prayers in front of a ***shiva*** idol. This is an outcome of ***Bhakti yoga and Gyan yoga.*** The key is Yoga!

CONCLUSION

Without Yoga, knowledge remains fragmented, dry, and in specialized compartments. It also takes much longer for the resolution of a problem. Yoga can open up the secrets locked into the vaults of Vedic ***mantras***. Mere knowledge of mathematics and Sanskrit alone cannot unlock especially the mathematical contents of the ***mantras***.

The construction of such Vedic ***sutras*** are beyond the capabilities of scholars trained in the modern education system. Unless and until, Yoga leading to some ***siddhi*** is integrated in the total learning process, mathematics in the Vedas will continue to remain hidden. What one sees on the surface are like pebbles on a sea shore; precious gems and pearls are buried underneath the sea.

REFERENCES

1. Tirath, Bharath Krishen, ***Vedic mathematics*** Motilal Banarsidass, Delhi, 1965.
2. Beckmann, Petr, ***A History of Pi***, Golan Press, Boulder, Colorado, 1977.
3. Shastri, Kashi Ram, Regiment Bazar, Ambala Cantt, Haryana, 1986.

November 22, 1986

41. HISTORY & PHILOSOPHY OF SCIENCE: A VEDIC PERSPECTIVE

{**Note, January 2021:** This article was published in the Proceedings of the **First Seminar on History & Philosophy of Science** held on the campus of Guru Nanak Dev University, Amritsar, India, during March 2-4, 1987. It is edited for this book, but no significant changes have been made.]

History of a nation, place, thing, or of an idea develops only when it is given some social recognition by a particular society. For instance, ancient Greek society dating back to 800 B.C. recognized the importance of science and mathematics much more than contemporary India has done it. In Greece, one can find a wealth of systematic material on science and mathematics from the ancient times as compared to very little in India. There is a historical reason for this difference.

Even the organization of this First seminar taking place after forty years of independence points out to the state of science in India. Presently, science, both theoretical and experimental, is being transplanted in India rather than indigenously cultured. In the USA, universities, like Indiana University, have departments of History and Philosophy of Science, which are nearly 100 years old. In contrast, India has a long way to go in such an intellectual direction.!

Philosophy of a person, or of a discipline evolves through various events, scrutinies and applications. It comes out like cream on the top that some tangible and unifying statements can be made out. For example, one says: Non-violence in the political arena is Gandhi's philosophy of action. Philosophy of art lies in its expressiveness. Therefore, one has to dwell deeper into the Vedas in search of these answers when it comes to the history and philosophy of science.

Philosophy of science emerges when confronted by a problem, scientists throw up their arms for a while. One can notice it in all the statements of great scientific minds when they are not able to resolve a particular problem, or find themselves facing a collision between a theoretical prediction and an experimental observation [1].

VEDIC LITERATURE

The Vedas are acknowledged all over the civilized world as the repository of knowledge, but in a highly distilled form. The four **Vedas** along with their four ***upvedas*** and six ***vedangas*** form the nucleus of Vedic literature. The four Vedas are ***Rig Veda***, ***Yajur Veda***, ***Atharva Veda*** and ***Sam Veda***. The four Upvedas are ***Ayurveda***, ***Gandharvaveda***, ***Dhanurveda*** and ***Sthapatya Veda***. The six ***Vedangas*** are **grammar**, **prosody, astrology**, **lexicography**, **etymology** and **rituals** with proper pronunciation. They are essential for the proper understanding and interpretation of the Vedas. The verses in the Vedas contain knowledge in the seed form, and hence their literal interpretation would be generally unmeaningful, as they are also sound based. Therefore, one is bound to wonder as to what place has been accorded to mathematics/ mathematicians and science/scientists in the Vedas.

HISTORY OF SCIENCE IN THE VEDAS

During the last millennium, the Hindu India lost all its intellectual traditions including its history. There are a few stray and scattered written records - like the ones written by two Chinese scholars in the 6th century AD. The modern history of India is colonial - Muslims followed by Europeans. On a personal note, I recall the only college level history book in India was first put together by Anderson. It was constructed out of the diaries that Anderson kept posted as an officer of Indian Civil Service. By and large, history in India is interwoven in its folklore.

Early European scholars along with a few Indians have tried to establish various chronological accounts from the Vedas. However, they all proved to be inconsistent. One can refer to the work of Pandit Raghunandan Sharma [3], first published nearly 50 years ago, and recently republished. Reason for their being incorrect is that they have tried to give modern dictionary meanings to the words in the Vedas. Therefore, as it stands, no one has been able to formulate any history of science and mathematics pertaining to the Vedic period! Nonetheless, there is overwhelming evidence in the Vedas that science and mathematics were as fully developed as other aspects of human knowledge like art, music and literature.

Morris Kline, in his monumental work [2] on the history of mathematical thought, simply brushed aside any reference to the Vedic period of India.

He is justified to some extent. The Vedas contain no specific names of mathematicians, physicists, or engineers per se. This does not mean there is no reference to topics of science and mathematics in the Vedas. Vaidyanath Shastri's book ***Sciences in the Vedas*** [4] contain precise references to Vedic ***mantras*** on various topics in science and mathematics. Recently, efforts of Maharishi Mahesh Yogi have been highly successful in linking several modern theories of physical sciences with specific ***mantras*** in the Vedas.

PHILOSOPHY OF SCIENCE IN THE VEDAS

Despite copious scientific references in the Vedas, Vedic literature has not put science relatively on a high pedestal. The reason is very simple. History of modern science clearly tells that science is basically utilitarian. It has enhanced the number of objects, which a man can possess and consume. It thus caters to the material welfare of man. Science pushes an individual to conform to organizations whether in production, or urban lifestyles.

In contrast to the above observations, one finds that in the Vedas, the emphasis is on the total upliftment of man. Spiritual growth takes the highest priority. Science does not recognize spirit or soul as such. Nobel Laureate, George Wald (1965, in Biochemistry) is the first top scientist who incorporated consciousness and the Vedic view in science, for which he was ostracized by the scientific community. He was also a keynote speaker in the conference on ***Science and Religion*** held in Bombay in January 1986.

Recent paradoxes in basic sciences are forcing scientists to look deeper into the foundations of science. Foundations of all modern sciences converge in mathematics. Since **Gödel's Incompleteness Theorem** and **Heisenberg's Principle of Indeterminacy** (both proved/discovered in Germany around 1930), foundations of mathematics and sciences have ever been shiftier. No matter how sophisticated a theory is developed to explain old paradoxes, it runs into new paradoxes eventually.

In the Vedic approach, the mind is trained holistically (***yogic***). Yoga is the union of an individual's consciousness with the universal superconsciousness. **Faith thus turns out as transcendental logic**. It is not anti-logic! There will remain a gap between logic and faith as long as one continues to stick to the Newtonian scientific view of uniform demonstrability. However, the sages or the ***rishis*** of Vedic literature have a unifying approach in the whole learning process. The very composition of verses in the Vedas epitomizes the unparalleled development of the human mind, that some take them as a work of divine revelation.

There are numerous instances in the Vedas when a particular ***shloka/mantra*** is found to have several distinct meanings including one highly scientific. Shankaracharya Bharati Krishen Tirath astounded the mathematical world by distilling sixteen mathematical aphorisms (***sutras***) out of the Vedas, which as they are, cannot be found in the present recensions of the Vedas [5]. He believed: True realization (enlightenment) is actual visualization, therefore, it has to have a mathematical form too! In conclusion, according to the Vedas, any philosophy of modern science has to encompass the total philosophy of living.

REFERENCES

1. Capra, Fritjof: ***The Tao of Physics***, Shambhala Publications. Inc., 1975.

2. Kline, Morris: ***Mathematical Thought from Ancient to Modern Times***, Oxford University Press,1972.

3. Sharma, Raghunandan: ***Vedic Sampatti*** (Hindi), ***Samarpan shodh Sansthan***, 1984.

4. Shastri, Vaidyanath: ***Sciences in the Vedas***, Sarvadeshik Arya Pratinidhi Sabha, 1970.

5. Tirath, Bharati krishen (Shankaracharya): ***Vedic Mathematics***, Motilal Banarsidass, 1970.

December, 1987

42. FERMAT'S LAST THEOREM & MATHEMATICS IN THE VEDAS

[**Note**. The manuscript of this paper was discovered in a folder amongst other manuscripts after the publication of the second volume of the mathematical reflections. Since they were typed on the earliest version of Apple PC, there was no way of revising them on the WORD, which did not exist then. Hence, they were re-typed by an office secretary by squeezing them in between more important chores. Psychologically, I find it so difficult to type my old material myself. However, some of its ideas have been included in articles published in Volume II. However, it is worth including in this volume.]

BACKGROUND

The book, ***Vedic Mathematics*** by the late Shankaracharya Bharati Krishna Tiratha [8] has been in print for over two decades. To the best of my knowledge, no one has written a detailed scholarly review of this book as Swami Satya Prakash Saraswati [6] has done in the very first issue of ***The Journal of International Dayananda Veda-Peetha.*** Over and over again, he reiterates that all the sixteen ***sutras*** and their corollaries, as discovered by Shankaracharya, have no connection whatsoever with the Vedas, Vedangas, Upvedas, or their Parisistas as stated by Shankaracharya. However, Swamiji has given full credit to the genius of Shankaracharya in discovering these ***sutras*** as a feat of great intuition.

After reading this review, two parallel thoughts struck my mind; one, from the popular folk tales of India and the other from the world of mathematics. In the first story, an aging wealthy farmer remained worried that on account of the laziness of his sons who never learned the habits of hard work, the family fortune would soon disappear. After his death, the sons found a note in the will stating that he had hidden a pot of gold somewhere in his field. The greed for gold energized the lazy sons, and they started digging the field from one end to the other. No gold pot was found though. But after the rains, the crop was bountiful as the soil was very well prepared in the process of the search for gold. It was then that the sons realized how their father wanted to teach them a lesson on the virtue of physical labor.

The second story is that of famous **Fermat's Last Theorem,** which states, that there are no positive integer solutions of the Diophantine equation, $x^n + y^n = z^n$, for integer n > 2. Fermat (1601 – 1665), a number theorist of the seventeenth century had noted (in 1637) in the margin of his copy of the works of Diophantus, "I have discovered a truly remarkable proof which this margin is too small to contain." For the last three hundred years, the best mathematicians all over the world have been trying to prove it, or disprove it. Mathematicians are convinced that Fermat did not have a proof of this theorem. Fermat proved it for a special case of n=4. It took nearly 100 more years and the mind of the greatest number theorist Euler (1707 – 1783), to prove this theorem for n=3. However, the general proof remains elusive. Nevertheless, the interest in the problem continues unabated. New Mexico State University has invited the top mathematicians to a Seminar on Fermat's Last Theorem during December 1988.

A MATHEMATICAL MORAL

What is the point of these stories? The genuine efforts for the search of truth are amply rewarded. In the folk tale, the sons eventually became industrious and every year their fields yielded gold in the form of bumper crops. In the second case, as mathematicians are aware, few other problems have generated so many new ideas and areas in mathematics as this theorem has already done. Algebraic and analytic number theory are two major areas of mathematics opened up by this problem. One can refer to [4] for some details related to number theory. This problem also had an impact on numerical analysis/methods and computer software and hardware alike when this theorem was tested for up to ten digit values of n. No less important is a conviction of many mathematicians that this theorem may not be ever proved or disproved. This had been due to another far-reaching result known as the ***Incompleteness Theorem*** in mathematical logic, proved by Gödel in 1931. It states in simple language, that in any algebraic system there exist statements, which cannot be proved or disproved. Fermat's Last Theorem may belong to this category!

MATHEMATICS IN THE VEDAS

All this prefatory detail is essential to my analysis on the nature of mathematics in the Vedas. The Vedas are not the books of mathematics, as one knows them today. There are several mantras in the Vedas, which enumerate various arithmetic progressions. See references [5, 7]. The most pertinent question is, what is the purpose of these mantras in the Vedas? If the only purpose of these mantras is the superficial scattering of trivial mathematical expressions here and there, then by no means are the Vedas to be acclaimed as divinely inspired work. That is why some European Indologists have even called Vedic hymns the songs of the shepherds!

VEDIC SANSKRIT

It must be emphasized that mathematics in the Vedas is not written in the manner that a person can simply deduce it by knowing some nuances of the language. Knowledge in the Vedas is in the seed form - highly compact and rarified, and composed in verses which are sound based. That is why some mantras are nearly repeated several times, and in more than one Veda. The oral tradition for the preservation of Vedic knowledge coming down from millennia, though still existent, was codified in a written form in course of time. However, it must have taken a hundred years for the present form to crystallize in Sanskrit language at the pinnacle for its syntax and semantics. On the top, mathematics being a language of science becomes a language within a language!

An analogy of Sanskrit language with a locker is very appropriate. One keeps one's most valuable objects in a locker placed in the innermost and strongest chamber of the house or bank, and its access key is known to only a few persons. To preserve the authenticity of knowledge, Sanskrit continued to be highly sophisticated and structured relative to other languages of India. Appropriately, it is called the language of the ***devas***, meaning thereby that it was beyond the intellectual grasp of the ordinary scholars not to mention the talk of the common man. It is not out of place to add here that it is absence of such a developed language that all other great civilizations of the past like Inca, Babylonian, Egyptian, Greek, Roman, and Maya have left either no trace their greatness in any written work except in some archeological ruins, or relatively small as compared with that of India.

STUDY OF THE VEDAS

There is a strict methodology for the proper study of the Vedas. Its comprehensive methodology has been given by Swami Dayanand [3]. It spans a period of intensive study for at least 25 years! Starting from the phonetics by Pannini, one studies grammar including ***Ashtadhyayi,***

Dhatupatha, ***Ganapatha***, ***Unadikosha***, and ***Mahabhashya***; ***Nighantu*** and ***Nirukta***, books on Vedic vocabulary and philology, and then ***Chhandograntha*** (Prosody). The study of the Vedas is still down the line. One is directed to study ***Manu Smriti***, ***Valmiki Ramayan***, ***Viduraniti*** and portions of ***Mahabharata***, six ***shastras*** and ten ***upanishads***. After all this study, one is ready for the proper study of the four Vedas and their four ***brahmins***, which according to Swami Dayanand, alone takes six years. Afterwards, one has further to go into the study of the four ***upavedas***, and finally into the ***jyotish shastra***.

CURRENT STATUS

I am not aware of any university and college in India or abroad where such a comprehensive schedule of the study of Vedic literature is prescribed. Assuming there are some isolated scholars of Sanskrit, but they would have little knowledge of modern mathematics and science. In order to successfully comprehend the mathematical meanings of the mantras in the Vedas, one has to be a great scholar of both Sanskrit and mathematics. Unfortunately, such a combination has been non-existent in Indian curricula for more than a century. At present, those who study Sanskrit in schools and colleges don't go near science and mathematics, and vice-versa. Since Lord Macaulay brought the so-called "education reform" in India around 1830, many generations have passed without producing any great scholars in science and Sanskrit. Ironically, today the greatest opposition and even ridicule towards mathematics and sciences in the Vedas, come from the academicians. Well, wasn't it the whole cultural conspiracy of the British through Macaulay's education reform?

YOGA AND MATHEMATICS

Knowledge of Sanskrit and mathematics are necessary, but not sufficient conditions for figuring out the essentials of mathematics in the Vedas! The third and the most important ingredient is the power of yoga. It is often said that verses in the Vedas have yogic meanings and one must have a meditative and highly contemplative mind to be able to fractionally distillate its shades of meanings. What does it really mean? It is the harnessing of holistic knowledge from its very source by tuning the mind directly with that source. India has so much lore and literature on the techniques for training the mind. ***Vigyan Bhhairava sahastra*** is the earliest work on this subject coming down to us as a part of India's great heritage. This is exactly what Shankaracharya did when he went into the forests of Sringeri and did ***tapasya*** for eight years. He had an exceptionally brilliant mind and had gone through formal training of modern higher education in mathematics and Sanskrit.

His biographers tell that mathematics was his passion, and exposition of Hindu religion and philosophy his way of life as the Shankaracharya of Puri. He firmly believed that **Self-Realization** is **Actual Visualization**. If that is what true knowledge is, then one of its forms has to be mathematical too. That is how mathematics finds its place in the Vedas, which expound the laws of universe and human mind! The ancient Greek philosophers of Pythagorean School were more specific; they believed that natural numbers describe all the laws of the universe. Perhaps these superficial references to the numbers in many mantras in the Vedas signify more than that. But it will require the genius of several Shankaracharyas to completely decode them! The only paper [1] on this theme was presented during the ***First World Yoga Conference*** held in New Delhi during Dec. 29-30, 1986.

One can understand the multiplicity of meanings of the Vedic mantras by comparing them to the modern audio cassettes whose tapes can have up to four different pieces of information recorded on the same side. However, one can retrieve it only after bringing the magnetic head in precise contact with the right segment of the tape. ***rishis*** are considered as the seers (***drashtas***) of mantras, and not their makers.

CONCLUDING REMARKS

In modern terminology, any knowledge gained through intuition does not mean that one gets it without ever consciously thinking about it. On the contrary, Western psychology now also asserts that intuition is based on past experiences, and is a process of pulling cues from the unconscious mind and bringing them to the surface. At times, a controversy is created when the inventor is asked to explain the genesis of his invention, which actually belongs to the domain of Meta logic. The greater the invention, the harder it is for other experts and the general public to accept an explanation. The work of Ramanujan (1887-1919) baffled the great contemporary mathematicians of the caliber of Hardy and Littlewood. When Polya (1887-1985) first saw Ramanujan's work in Cambridge, putting it aside, he remarked that if he were to try to understand it, then he would not be able to do any mathematics of his own.

According to his biographers Aiyer and Rao, Ramanujan attributed his discoveries of mathematical formulas to the goddess ***namagiri*** of namakkal in his dreams. It is a remarkable fact that frequently, on rising from the bed, he would note down results and rapidly verify them, though he was not always able to supply a rigorous proof. Some people particularly of Western thinking may even find it absurd. However, it is not. Ramanujan through ***Bhakti Yoga*** penetrated regions of mathematics, which were hitherto unknown. After 50 years, a renewed interest in his work, for its applications in computing, is projecting Ramanujan as a genius of unparalleled order in the eyes of the public. One has to see a recent movie produced by Nova on his life and work, ***THE MAN WHO LOVED NUMBERS*** with commentary by the US number theorist George Andrews. The update on his work will be a topic of panel discussion during the joint meeting of the American Mathematical Society and the Mathematical Association of America at Phoenix, Arizona in Jan 1989.

If the SUTRAS did not have the remotest connection with the Vedas, then why did Shankaracharya give the whole credit of their origin to the Vedas in no uncertain manner? Why did he not take the whole credit for himself? One thing is clear to me, as a Hindu, that Shankaracharya did not lie, or distort the facts as he **SAW** them. With this premise, I have posed seven open questions about mathematics in the Vedas [2]. The very first question is the tracing of these SUTRAS in the Vedas. If the source is not in the Vedas, then where is it? Like Fermat's Last Theorem, all stand to gain in the process of finding an answer to this question. If history is any guide, then in my opinion the resolution of this question is not within the means and abilities of one individual, one institution, or even one generation. It will require the collective will of the Vedic scholars in India and abroad along with the artificial intelligence of the latest computers to unearth the mysteries of Vedic hymns.

ACKNOWLEDGEMENTS
Thanks to Dr. R. D. Saxena for his assistance in locating some material, and Dr. David W. Emerson, Dean, College of Science and Mathematics for his suggestions in improving the effectiveness of the ideas.

PS: March, 2016: Dr Saxena died in 1994, David Emerson in 2012. Andrew Wiles (1953 – Present), a British mathematician of Institute of Advanced Studies proved Fermat's ***Last Theorem*** in 1995. In 358 years, it has involved generations of mathematicians and produced hundreds of doctoral dissertations.

REFERENCES

1. Bhatnagar, Satish C. ***Mathematics in the Vedas and Yoga***. Yoga Mandir, pp 14-15, Vol 13 (7), July, 1987.

2. -----------------, ***Seven Open Questions about Mathematics in the Vedas***. Yoga Mandir pp 41-42, Vol 14(7), July 1988.

3. Dayanand, Swami. ***Satyarath Prakash***, First English Translation (1906) pp 65-70. Published by Arya Samaj Mandir, Madras, 1932.

4. Koblitz, N. (Ed.), ***Number Theory related to Fermat's Last Theorem***. Progress in Mathematics, Vol 26, Birkhauser Boston, 1987.

5. Satyakam, ***The Holy Vedas***, pp 416-420. Clarion Books, 1983.

6. Satya Prakash, ***Vedic Mathematics-A Review***. The Journal of International Dayanand Veda Peetha, pp 123-127, Vol 1(1), 1988.

7. Shastri, Vaidyanath, ***Sciences in the Vedas***, pp 21-42. Sarvadeshik Arya Pratinidhi Sabha, 1970.

8. Tirath, Shankaracharya Bharati Krishna, ***Vedic Mathematics***. Motilal Banarsidass, 1965.

November, 1988

43. HISTORY-PATIALA-MATHEMATICS

[**Note**: A reflective note to the organizers of the International Conference on the Advancement of Mathematical Sciences, at GSSDGS Khalsa College, Patiala, India - during March 19-21, 2015.]

An invitational email about a mathematics conference in Khalsa College Patiala immediately stirred up a lot of memories in my mind. For an historian, the name General Shivdev Singh Diwan Gurbachan Singh (GSSDGS) Khalsa College, Patiala is a new name of the plain Khalsa College. I have known the College since the 1960s when it was run in shed-like buildings, and professors' salaries used to be in arrears for months. Two Chahal brothers, acquainted with me, taught there. The one who taught English died a few years ago, but the other one, Harpal who taught political science moved to Yuba City (USA), which I have described in my book, ***Vectors in History*** (2012) as a citadel of ***Punjabiyat*** in the US.

I grew up in Bathinda (which still has not come out of the image of '***Via Bhatinda'****)* in the 1950s when Patiala was a little 'paradise' in my imagination. I used to say that once I got a job in Patiala, I would spend the rest of my life there. Lo and behold, when Multanimal Modi College, Patiala opened up in July 1967, I became the one-man Head of the Mathematics Department. But the birthing pains of this new institution created such an administrative hurricane that I had to leave not only my beloved city, but also the state, and even India in 1968 (which I was adamant on never leaving!).

My two older children were born in Patiala. We had a nice house in Ragho Majra, 200 yards away from the College. My wife completed her MA in Philosophy from Mahindra College Patiala, where one BK Bhalla, transferred from Govt. Rajindra College, Bhatinda, was one of the math professors. My late friend, Vikrma Ditya Singh was the first Professor and Head of the newly started Punjabi University, Patiala (1962). I cannot remember the names of any math professors at Khalsa College. However, one Mrs. Saxena taught math in the Government College for Women!

During my brief stay in Patiala, I did try to organize a mathematics group in Patiala.

For your information, I have known a conference keynote speaker, IBS Passi, since 1959; when I started MA Part I, he was ahead in Part II. We still remain in touch with each other through my reflective writings. From an academic point of view, the only living acquaintance of mine left out in Patiala is Surindra Lal, my 1967 physics colleague, who retired as Principal/Director of Modi College. He is now a leading author of the college physics textbooks.

Apart from sharing a trip down my Patiala memory lanes, I would urge the organizers to include a session on the ***History of Mathematics in Punjab***, one of my ongoing research areas. In fact, I would challenge the organizers to focus on the contributions of the Sikh mathematicians. The world knows the Sikhs for their chivalry and bravery. According to one online source, the Sikhs are only 2% of India's population, but they make up 8% of the present army/defense forces, and are far more dominant in athletics. As a corollary of the great Sikh Empire under Maharaja Ranjit Singh, the Sikhs are famous in politics and businesses even outside India. The number of Sikhs in arts and literature is proportional to their numbers.

However, when it comes to the hard-core mathematical sciences, the number of Sikh researchers dwindles down, and it goes to zero, in a limit, when it comes to mathematics. Here, **by researchers I mean beyond PhDs and full professorships**. The conference must bring to light the Sikh state in mathematics. A conservation principle explains this gap partly. For instance, if Sikhs in the army are four times the Sikh percentage, then mathematically there exists at least one discipline, including mathematics, where the percentage of the Sikhs has to be at less than one-tenth of one percent. But why Mathematics?

This 'conjecture' may serve as a Null hypothesis in analyzing data, which I have not been able to find or collect. Thus, it is an open research problem both in the areas of Mathematics Education and in the History of Mathematics. That may also be set as a benchmark for this conference. In addition, what is worth examining is the role of the Sikh religious ethos and political tenets that play out in the development of mathematical mind-sets. It is another angle, which no one has examined in a popular area of Ethno-Mathematics.

On a related matter, next April, I am organizing a session on the History of Mathematics. The details are on the AMS website on its Western Section meeting at the UNLV campus. Here is description, which you may like to publicize and/or include in your conference program:

"***In mathematical jargon, the objectives of this session are to explore the necessary and sufficient conditions for the development and flourishing of mathematics. From a cursory examination of history, it is clear that these conditions are functions of at least three factors; namely, political systems, educational models, and organized religions. These are structured institutions and are reflected in the growth of mathematics, which is a super organized discipline.***

At any point in history, if the entire world is surveyed mathematically, then one finds a few green spots (countries) where math thrives (homes of the Fields Medalists and Abel Prize winners, etc.). Next come the brown spots where fundamental math research is not on the horizon. Finally, one sees swaths of barren spots, where math is stressing to sprout. Thus, papers are invited where the focus may be on an individual, society, institution, or ideology in the present or at any time and place in the past for the validation of these objectives. This session cuts across the traditional boundaries of history, politics, psychology, sociology, science and religions."

I wish the conference organizers the best of success, with the hope that this event does not turn out to be like a one night stand. Such conferences should be held every two-four years in order to nurture mathematical minds in Punjab. Though late in knowing about it, still, I would be glad to assist or help in any manner. Feel free to contact me.

Finally, if there are good undergraduate mathematics students in Khalsa College who also want to have an experience of US life, then encourage them to apply to UNLV. The chances of graduate assistantships are high for a PhD program, though a few are granted to foreign students for MS too. Our PhD program is young – only ten years old. The students may be encouraged to contact me. Regards.

Feb. 12, 2015

COMMENTS

Esteemed Professor, Dr. **Satish C. Bhatnagar**: We are really feeling pleased to read the mail sent by you. We are proud of a ***Patialvi*** being at a reputed position in University of Nevada, Las Vegas. Sir, I have spent almost 5 years in company of Prof. Inderjit Singh Chahal. I joined the college in 1995. Since then, I have seen a lot of changes in this college. Specially, when our present Principal Dr. Dharminder Singh Ubha joined the college as Principal in 2008. You will really be astonished knowing that the strength of the college has increased from 1200 to 6000 in the last 6 years. Recently, the Minister of higher education declared that the college will be given the status of university soon and the name of the university will be Sri Guru Teg Bahadur University. On behalf of the college staff and Principal, I thank you once again for giving us golden memories of our college.

Also, I invite you as Chief guest on the valedictory function of ICAMS 2015 to be held in the college from 19-21 March, 2015, if you agree to be a part of this International event. Waiting for your positive reply, Regards; Prof. **Gurmeet Singh**; Convenor ICAMS 2015

44. RP BAMBAH - AS I RECALL

The celebration of the decem years of a distinguished professor by one's colleagues, friends and students is a laudable western scholarly tradition. It could happen during one's active years of service, or after retirement. In the US, there are all sorts of ways of organizing such events - like, seminars, symposia, colloquia, small conferences in exotic locales, and publishing memorabilia, etc. Thus, new benchmarks of scholarship are set out for the new generations. It often begins with the completion of 60 years; but 70th is the most popular anniversary. I have seen research papers dedicated to the 80th anniversary too. However, by the 90th anniversary, either the celebratee is physically and/or mentally gone hiking, or the number of celebrators have dwindled away. Personally, I have yet to be a part of any kind of 100th anniversary.

The history of India is so checkered that the indigenous intellectual traditions have not yet taken institutional roots. The main reason being - the intellectual traditions flourish only in an atmosphere of absolute political freedom. The Hindus, once constituting nearly 100% of India's population, lost Hindustan, their sole homeland, to hundreds of Muslim and Christian bands of invaders - from next-door Afghanistan to far-off Turkey, to the regions of Central Asia, and literally from every nook and cranny of Europe. They decimated great Hindu centers of learning, burnt down lofty libraries, and demolished grandiose temples that also preserved rare manuscripts. Above all, they all terrorized the populace by inflicting regular holocausts that changed the religious landscape of India forever. It is depressing to recall the last millennium.

India is politically free today, but the mindset of the Hindus is shackled in the deepest fears both in their individual and collective psyche - physical, mental and spiritual. It surfaces up in their beliefs and practices of convoluted secularism, equality, peace and tolerance in every facet of public life. Consequently, intellectual traditions in India have a long way to go before they would sprout. However, I do remain optimistic about them.

This train of thought struck my mind as I recalled that RP Bambah (Ram Prakash Bambah), distinguished mathematician, renowned Vice Chancellor

(1985-91, equivalent to president of a US university) of Panjab University (PU), Chandigarh, a prominent humanist, and Padma Bhushan awardee, who turns 90 on Sept. 30. A few years ago, he stepped down as one of the five Trustees of the ***Tribune***, an English daily newspaper of Punjab (established in 1881). The list of Bambah's honors and awards is so long that it is safe to bet that there is no mathematical sciences award of India that has not been bestowed upon him.

I first came to know of RP Bambah as his student in July 1959 when he taught the ***Statics and Dynamics*** paper (nearly a two-semester course sequence) of the MA Part I (1959-60) for a week or two, as a substitute instructor. This subject was miles away from his area of expertise. Instead of saying No to the Department Head, he faced a few embarrassing moments in front of the students while heaving and fumbling during his lectures. Nevertheless, I admired his guts, as this eventually contributed to my forte in teaching. Now, I proudly tell you that I have taught a record number of nearly 75 different mathematics courses and seminars to various clientele at the University of Nevada Las Vegas alone.

In my circle of super senior acquaintances, Bambah is a rare exception - in the sense that all his faculties are still intact even at this ripe age. I was so delighted to see him on June 27, 2015, when he drove up all by himself to a gathering of friends and relatives that I had organized in the Panjab University Alumni Center, Chandigarh during my last visit to India.

I have come to know that the PU Mathematics Department is organizing a one-day seminar in the honour of RP Bambah's 90th anniversary. Generally, dry research papers are presented in clinical manners. Nevertheless, it has stirred memories of my forlorn academic path. Had I followed a linear path, I may have done my PhD in Number Theory, the field of Bambah's research. A particular classroom moment is worth recalling here; it was during the MA Part II (1960-61) and Bambah was teaching his favourite paper on **Number Theory**. Over the course of a week of lecturing, when he was detailing out a proof of the ***Prime Number Theorem***, I suddenly observed a short cut. It prompted Bambah to remark, "How come Hardy missed it?". He was following the classic number theory textbook by Hardy and Wright.

Bambah's remark was one of the high points of my student days. Soon after the results of the MA Part II were announced, HR Gupta (1902-1988), the Weierstrass of India, and the then Head of the Department, offered me one of the two research scholarships for PhD studies. It was very prestigious then, but I declined it right away in favor of going for the elite Indian Administrative Services!

One of the popular legends of RP Bambah is his scoring 400 marks out of 400 in his MA (1945) exams from PU Lahore. It could only be equaled by the year 1947, as afterwards, PU was bifurcated between India and Pakistan. Soon after independence, with two additional papers, the aggregate became 600. It was 800 during my era with eight papers, and now it is 1000 with ten. According to one source, Abdul Salam (29 Jan 1926-1996), the Pakistani theoretical physicist and the 1979-Nobel Laureate (joint) and RP Bambah went to the same school and college. Salam also scored 400 points in his MA (Mathematics) exams (not double-checked). After finishing PhD from Cambridge (1950), Bambah joined PU Chandigarh straight as a reader (equivalent to associate professor) in 1952 and was promoted to the rank of (full) professor in 1957 - the youngest mathematics professor of his times - at the age of 32.

Often people speculate that life's goodies came to Bambah rather easily in a highly corrupt and competitive job environment in India. In fact, Bambah's early upbringing and mine are interestingly similar. Family wise, our fathers were railway employees and mothers nurtured the families exclusively. Our social values were imbibed and instilled by studying in the Arya Samaj schools. We refused to take any dowries, and married girls of our choice and heart. Some of these tidbits have recently been shared by Mohinder Singh Cheema (1929 -), a five star mathematics product of PU, who has been living a post professorial life in Tucson, Arizona.

However, there is a 180-degree difference between our lives, when it comes to connections or networking. Bambah and his wife, Soudamini (died in 2011) met on a ship in 1950 while returning to India from the UK. Bambah still has the handsome looks, but had no romantic streaks, as we students used to joke about. He still wears harried administrative looks. Anyway, they were married soon after arrival in India. Mrs. Bambah earned all the accolades on her merits. She was the daughter of Dr PK Parija (1891-1978), an internationally known botanist, a Cambridge don of the 1920s, the first

Vice Chancellor of Utkal University in Bhubaneswar, and a member of the Odisha Legislative Assembly. With Bambah's sterling academic and administrative record and his "Parija connections", there just had to be a resonance effect. Sometimes, the doors may have opened up for him even without a knock.

Three months ago, when I saw Bambah after some 30 years, he still had a full head of hair - majestically graying, the same facial contours - no wrinkles! In walking and talking, he did not miss a beat. I could not help asking him if he had ever exercised in life. Promptly, he said, "Never!" I did not have to ask further about any secrets of his healthful longevity. On my part, there is hardly a day when I do not spend an hour in yogercises, calisthenics, long walks, and/or bench presses (horizontal) - almost addictive! Of course, my college days in Bathinda and that of Bambah in Lahore, an epicenter of Indian nationalism before 1947, exposed us to the opposite extremes of life.

Nevertheless, it is still too early to measure the legacy of RP Bambah. But, I can't help adding some lines on his character and integrity. His tenure as Vice Chancellor was amidst the darkest period in Punjab's history when the Sikh militancy, unrest, or insurgency was at its peak. It was fully exploited, supported and fomented by Pakistan, which was waiting to take revenge on India for the breakup of Pakistan in 1971 and the creation of Bangladesh. On politicized academic issues, Bambah resigned from PU, but the Syndicate did not accept his resignation, and went on record by granting him indefinite leave of absence because of his exceptional services. It speaks of his principles and courage. For a number of reasons, he did leave India however, and joined the Ohio State University, Columbus, which had always been open for him due to his scholarship.

Thank you, Bambah Sahib for the memories and inspiration of a lifetime. May you continue to remain active enough so that I may witness your centennial celebrations! By the way, the legendary columnist, Khushwant Singh (1915-2014), a multi-dimensional intellectual of India and Padma Vibhushan awardee, had missed his centennial by only a few months. This may be one last record for you to break!

September17, 2015

COMMENTS

Great professor and great student like you. **Vinay Shirodkar**

I am glad for the privilege to have met Prof. Bambah at Ohio State University in the early 1960s. Believe it or not, he agreed to come to our small nest for dinner cooked by Joginder. **Harbans Singh Bhola**

Hi Satish: After a long hiatus, you now have a 'reflection' on your favorite professor who is still active and productive in his ripe old age. My hearty congrats to him and your sentimentally charged piece that demonstrates your admiration of him and his contributions. **SS Moorty**

Dear Sir, Thank you for your write-up. I am taking liberty to forward it to the Editor, ***The Mathematics Student*** who would look into it and take appropriate decision in the context of RPB turning 90 on Sept, 2015 obviously with your concurrence. With warmth and greetings, **NK Thakare**

Thank you. Very interesting. **Shankar Raja**

Thanks for making me recollect yet one more time my wonderful time under the kind guidance of Prof. R.P. Bambah. Best regards, **Ajit Iqbal Singh**

Satish, A very nice personal tribute and some more history of you. Thank you for sharing. So, you could have been a number theorist, but instead chose government service. **Brian Winkel**

45. TIME TO HONOR SOM DATT CHOPRA!

(A Reflective Note to MD Sharma)

Your e-greetings triggered the sweet memories of the best couple of hours that I spent with you and your family last June. I was able to re-live my stay (1965-67) at Kurukshetra University (KU) campus, as you took me around. Naturally, it has jolted me into a concrete action - how about recognizing the best PhD dissertation/researcher from the KU Mathematics Department? Here are some suggestions from the top of my head.

1. The award is to be named after Prof Som Datt Chopra (1915-1987). During 1959-60, I was his student in MA Part I in the Mathematics Department of Panjab University (PU), Chandigarh. After getting PhD from Cambridge University, Chopra returned to PU, his professional home. When he was not promoted as a Reader (reason, academic politics), Chopra left PU in 1960. A new Regional Engineering College was starting up in Srinagar (Kashmir), and Chopra was hired as Professor and Head of the Mathematics Department. It eventually turned out to be good for both PU and KU. When KU was upgraded to a comprehensive university, Chopra joined KU in 1961, as the first Professor and Head of the Mathematics Department. He retired from KU in 1978.

Historical Sidebar: Jhangi Ram Arora and Joginder Pal were the only two students who finished MA in the first batch of 1963. In 1965, with Chopra's recommendation, I received a newly instituted UGC fellowship for doing PhD in Mathematical Seismology under his guidance. However, I left KU in 1967 without finishing the dissertation. To the best of my memory, no one had finished a PhD under Chopra before 1967. Nevertheless, Chopra had laid good research foundations in the Department. Looking back to the last 50 years, Som Datt Chopra's name with the best doctoral dissertation and research is justified. It is a tribute to his distinguished service of 17 years to the Department in particular, and KU at large.

2. **Evaluation Time Period**: 1961-1975. The year 1961 commemorates Chopra's arrival at KU and the year 1975 would give a good number of at least 10 dissertations of the researchers for evaluation.

3. **Impact**: Generally, the criteria are based on long-term impact of research–like, continuation in the research area, and diversification of the dissertation work in generating new modes, on the teaching and humanistic side. For instance, Chpora's English and relaxed enunciation rubbed off on me. **Sometimes, a good researcher is a good teacher too**. A KU researcher does not have to stop from being a good parent, spouse, friend, and a productive member of the society that nurtured him/her.

4. **Award**: I am putting Rs 10,000 for monetary award besides a citation conferred from the KU Vice Chancellor. I would encourage JR Arora, IBS Passi and Sarvajit Singh to pitch in. I will gladly match for the award to make it look good. I am not looking at it as a one-time award, but it could be awarded every 5-10 year. However, it all depends upon how this first award goes.

5. **Engagement**: Please share this write-up with the like-minded Math faculty, Department alumni, KU administrators, and whoever is enthusiastic to assist and work on a committee needed to refine the guidelines, evaluate dissertations and researchers, and finally select the first best (hopefully living) awardee. I can only assure a rewarding experience for everyone involved!

December 22, 2015

46. NAVIGATING MY STRAITS OF GEOMETRY

[**Note:** This Reflection is modified from an hour-long talk delivered on May 19, 2018 in the **Geometry Club** of Ahmedabad (India). It meets every Saturday afternoon under the aegis of the **Gujarat Ganit Mandal (GGM)**. After the talk, two geometry problems were discussed in a group of nearly 25 people.]

My fellow 'mathematical' friends – mathematical in the sense that we all chose to earn our bread and butter for the joy, propagation and service to mathematics. However, I am impressed with your love for mathematics in its various aspects. I always look forward to receiving and reading the monthly newsletter. The **Gujarat Ganit Mandal** (means Gujarat Mathematical Circle) may be the only such state organization of math lovers in North India.

Mathematicians, like GH Hardy and Bertrand Russell, have written about their inability to prove great theorems after their mid-40s. That is a reason that the Fields Medals are awarded to mathematicians under the age of 40 years. Nevertheless, we all can do enough mathematical activities to delay, if not completely ward off, the scourge of Alzheimer that strikes in old age.

This talk is a sequence of personal stories of my encounters with plane Euclidean Geometry, Analytic Geometry, Solid Geometry, Axiomatic Geometry, Differential Geometry; the teachers who taught me, and places where I learnt them; and so on, as much as I can mine geometries from the dusty vaults of my memory.

Also, the talk is set up in the format of a long-distance train journey in India, in which passengers embark and disembark according to their destinations. With some of them, we exchange tidbits of life, social, or political stories, but with a few of them, we even share our snacks and meals. I am talking of the time in India when trains were a major mode of travel. There was no kitchen pantry in trains then. We had our tiffin brought from home, and eating in a train was always enjoyable - a kind of picnic on the wheels. In other words, you would be able to identify yourself with some episodes in this talk.

High School and Geometry

Generally, initiation into mathematics is through arithmetic. The human awareness of numbers is innate. That is why counting numbers are called natural numbers. It is unlike the US school system, since the 1960s, dawn of New Math (or post sputnik era), where the notion of sets is introduced early on in elementary schools. Some educators argue that sets are more 'natural'. Students are able to understand the difference between numbers and numerals easily with associated geometry of figures. One could argue in favor of crude geometry of sets. Visual learning is quicker as, perhaps, evidenced by the TV marketing. On the balance, anything learnt easily is lost quickly too.

Naturally, most students resent variables, x, y, z etc. when algebra is introduced in the 7th or 8th grade. Personally, in a short time, I saw the power of algebra while doing classical problems of age and mixture, and I immensely enjoyed algebra afterwards. However, a quantum hurdle rose up in the 10th grade when Euclidean Geometry was introduced.

The memories of my two-week long frustration with the concept of proofs in Euclidean Geometry are still etched in my mind. I can recall that classroom, the teacher, Girdhari Lal, and finally the moment of enlightenment. I do not recall the textbook used for algebra or geometry. They were surely written by some Indian authors. The textbooks were lean and thin like the students then. In the US, there is a correlation between obesity and the size of a textbook. We bought the textbooks in the paperback and had them hardbound and then set them in a jacket of sturdy brown wrapping paper. Also, comes along is the ultimate reverence for textbooks, which for the Hindus are representatives of Sarasvati, the Goddess of knowledge.

College and Geometry

Algebra and Trigonometry were studied during the first two years (1955-57) of college, as a part of the F.Sc. exams administered by Panjab University at the end of the first two years. Ved Prakash Bansal was our math instructor, who openly said that his heart was not in teaching. Years later, I learnt that he retired as a principal of a government college! The other memory is of the trigonometry textbook written by Goverdhan Lal Bakhshi, a classmate of Professor Som Datt Chopra, the first Professor and Head of the Mathematics Department of the newly started Kurukshetra University, Kurukshetra. Bakhshi retired as the first Director of Public Instruction of Punjab. The other trig memory is that my cousin (student of a Kanpur college) and I challenged each other by sending trig problems through old-fashioned letters.

Looking back, I wonder at trigonometry, quite an established branch of undergraduate mathematics -stemming from the properties of a triangle. Of course, the circle is embedded in trigonometry, as trig functions are also called circular functions. The point is that no other geometric figure has generated such a body of mathematics. Am I right about it?

Calculus and Analytic Geometry formed the main syllabus in the last two years of the BA/BS degree in Punjab. The Plane Analytic geometry was a part of the regular syllabus and solid geometry of the Honors syllabus. Jagjeet Singh's textbook on plane analytic geometry was good, but the solid geometry one was written by Shanti Narayan, a nationalistic Arya Samaji, DAV College Delhi principal, self-taught mathematician, and the first Indian to write textbooks at the graduate level. Satish Chandra Deva was a newly hired lecturer of mathematics who was very conscientious in teaching. He married his student, who was the only female student in our class. It happened under very unusual circumstances.

My delightful moments of studying analytic geometry are associated with the power of its algebra over classical geometry. During the very first month, the lectures on the points of intersection of a line and a circle are so vivid in my memory. Visually, no common points of intersection corresponded to two imaginary points of intersection. The tangency corresponded to the two points being identical. Such mathematical encounters make mathematics extra-sensory - mathematics providing a sixth sense!

Euclidean Geometry of three dimension which is a part of one of the eleven books of Euclid was not studied in depth as geometry of 2-D. During the study of vector calculus, the applications of dot and cross product to the geometry of 3-D still fascinate me, particularly in the study of straight lines and planes. Beant Singh Grewal (?) was the first Punjabi to write a textbook on Vector Calculus.

In the US, at the undergraduate level, analytic geometries are relatively brushed aside. In high schools, geometry has become hybrid – a little bit of Euclidean and some axiomatic. As a parent, I saw it when my children were in high school. At UNLV, there is a one semester course, **College Geometry** (Math 480) designed for the students who are going to be teaching in high schools. This is one of the few courses that has escaped my teaching net.

Geometry during the Master's years

One of the four papers (a paper is equivalent to a 2-semester sequence) in MA Part I (1959-60) was on 3-D geometry. Everything about it remains unforgettable. The textbook, ***Analytical Geometry of the Conic Sections*** **by Edward Askwith** is remembered for its price of Rs 55, which was more than my monthly expenses of Rs 50, and was ¼ of my father's monthly salary, which supported my six other siblings too.

The instructor was Professor HR Gupta (1902-88), the Karl Weierstrass of India. Gupta taught the material off the textbook and in a style that was unorthodox for an exam oriented system. It created a storm of unrest amongst the students and I was one of the two student leaders. A paradox of life is how the twain meet. For the last 10-15 years, I have been recognizing the best student of MA Part I from the PU Mathematics Department in the name of Professor HR Gupta. It is a token of gratitude for his everlasting influence on my professional life – all because of his unique way of teaching geometry.

25% of this geometry paper was devoted to Differential Geometry that literally no one wanted to teach! I have no memories of its textbook, or of its instructor. Ironically, in order to learn Differential Geometry, I took this course in the summer of 1969 from a visiting assistant professor at Indiana University, Bloomington. However, it hardly turned out to be better than whatever was taught ten years earlier!

While digging the mines of geometries, I have faint memories of Professor SD Chopra lecturing on Spherical Geometry – the concept of a solid angle and sum of the angles of a triangle on a sphere.

Conclusion
In the US, the mathematical curriculum is reviewed periodically, so much so that it appears that topics are included and excluded like the trends in a fashion industry of hairdos, apparels and jewelry. The golden days of geometry are gone from the US classrooms. I no longer have an updated idea of it in Indian colleges and universities. Researchers go after certain geometrical properties. The fate of geometry is like that of special functions in higher mathematics.

Postscript: During the talk and at the end, I announced a kind of award for submitting the names of mathematicians from the state of Gujarat of today or yesterday - the national boundaries change every fifty years. Mathematicians include the knowers of geometry, ***yantra,*** astronomy, astrology etc. who had lived and died before 1192. ***Yantra*** is one of the three vertices of a triangle of divinity - ***mantra*** and ***tantra*** are the other two. An award of Rs 1000/mathematician for a maximum of three per person for an overall total of ten mathematicians. The deadline is August 15, 2018, and to be mailed to me; viabti1968@gmail.com. Not even a single submission was received - reflecting on narrow focus on mathematical interest.

May 20, 2018 (Ahmedabad/India)

47. SHADOWS OF SCIENCE AND *VIGYAN*

[**Note**: This reflection is modified from a one-hour invited lecture that I gave on June 14, 2018 at my alma mater (1959-61; MA), Department of Mathematics, Panjab University (PU), Chandigarh. This was organized under the auspices of the Chandigarh chapter of the ***Vigyan Parishad of India*** (Science Council of India) and sponsored by the local chapter of the ***National Academy of Science of India.***]

Background

During my India visits, I never miss walking the hallways and corridors of my high school, college, and university. On May 28, during such a casual visit to the PU Math Dept., an invitation was thrown at me to give a talk at the three-year old chapter of ***Vigyan Parishad***. On being informed about the goals of the ***Parishad,*** I readily accepted the invitation. Before starting on the topic, I paid a public tribute to Professor Hans Raj Gupta (1902-1988) for his impact on my professional life. He 'taught' me how to think creatively – something, which in essence, is unteachable! The talk was largely focused on the following five points.

1. Difference between Science and *Vigyan*

This is challenging. Since childhood, I have heard about their equivalency. However, one doesn't see the same outcomes in the corresponding societies - say, of science in the US and *vigyan* in India. This did trouble me. The Hindi/Sanskrit word, '***gyan***', which means 'knowledge' in English - both nearly equivalent. In Hindi grammar, the property of the prefix '***vi***' in front of '***gyan***' makes ***vigyan*** - means specialized knowledge, which could be like economics, music, dance, astronomy, math, etc. In mathematical jargon, science, as a set, is a proper subset of what comprises ***vigyan***. In other words, science + humanities + fine arts + engineering + social studies are almost equal to what ***vigyan*** stands for. Once this is agreed upon that the two terms are different, the challenge of giving this talk doubled up.

2. Foundations of Science and *Vigyan*

The foundation of modern science (since 1642, Newton's birth year) lies in experiments performed (hypothesis-observations-inferences) meticulously in the labs. The scientific thinking is that experiments performed under identical conditions produce identical outcomes. At this point, I also reminded the audience that the essence of mathematics is deductive thinking in which the reason for every step is given before it is questioned. That is why mathematics alone has theorems. The rest of the disciplines have theories. Yet, I do not mean to put mathematics at the top of any pyramid of all disciplines.

By the middle of the 20th century, the consequences of Einstein's Theory of Relativity, Heisenberg's Principle of Indeterminacy, and Gödel's Incompleteness Theorem in mathematics, and Principle of Butterfly Effect in Chaos Theory- they all shook the assumption that a scientific experiment can be repeated under identical conditions. Let me keep these great ideas in perspective. Einstein essentially proved that when the speeds are very high, then Newtonian physics is not adequate to explain certain physical phenomena. In other words, the universe which was deterministic in Newtonian physics, became probabilistic in Einstein's quantum physics.

Therefore, any science experiment is an end in itself, particularly, when an observer affects the very process of observing an object. It implies there is no absoluteness, objectivity and reality per se. This point of convergence is also called ***Maya*** – as derived in the metaphysics of ancient India. Despite such shifty foundations of science and mathematics, the progress in science and technology has been going on with leaps and bounds. What a miracle! It is due to collective mental constructs of people and the unboundedness of a single human mind.

A corresponding question on ***Vigyan*** is obviously not that simple. Here, I had to take a different approach and appeal to the knowledge, considered scientific by any modern norms, as mentioned in the Vedas, which alone have survived in their authenticity through millennia. When a society loses its political freedom, it loses everything - including its beautiful temples, institutions, women, and every treasure. Ultimately, the subjugated society stops to think freely! This is what has happened to the Hindus since 1192 - after the defeat of Prithviraj Chauhan. The sun started setting on them. The bottom line is that the political system of a land is the fountainhead of its science and ***vigyan***, and everything else.

3. Practitioners of Science and Vigyan

In all the four Vedas, there are many places dotted with scientific and mathematical information. How did this happen? They are compiled in a couple of books on **Vedic Mathematics** and **Vedic Science**. During my 2-year research (1986-88), I was shocked to discover that Indian scholars who knew science and mathematics were totally ignorant of the Vedic Sanskrit, and conversely, Sanskrit scholars went miles away from science and mathematics. It was the result of a devious plan under Macaulay's education system (good for the British!), which was imposed on India in the 1857 rebellion. The British Government gave scholarships for the study of Sanskrit but debarred them from government jobs. The college curriculum included British Literature, British History, British Economics, British Philosophy, and British Psychology. Once a UNLV professor asked me: "Is there any Indian philosophy?"

Let me make it clear that the study of science is very expensive. Moreover, the British created a perception that the superstitious Indians were not fit for the study of science. In fairness, this assertion is not all untrue. Even today, all kinds of superstitions are practiced in India. They are spelled out in various versions of the ***Laal Kitab*** (means ***Red Book***)- full of witchcraft and sold on street corners of India.

Science and mathematics are collaborative disciplines. The ancient Indian mind (***rishis***) harnessed the yogic powers to unravel the mysteries of the universe. The path of the yogic power is spelled out in preparing the body through the ***aasanas***, mind through ***pranayama***, and the union/***yog*** of the self with the Supreme Self through meditation. It is a limiting case of one's self-realization! This is not based upon my personal realization but as understood from literature and persons who seemed to have attained an elevated state of mind. I only had a glimpse of it when I took a 10-day ***Vipassana*** meditation course in Sep./Oct. 2007, and shall be repeating it next week (June 20 - July 01) in Dehradoon!

Nuclear energy is the iconic image of modern science, and life force/***praan shakti*** is the iconic image of ***vigyan*** - captured by the yogis/***rishis*** shown squatting in meditative postures. The yogis are walking nuclear reactors! The walls of the cores of nuclear reactors are 24"-48" thick and they are made of steel and concrete. Otherwise, with weak walls, the nuclear energy will shatter the cavity away. The same thing happens to the bodies of spiritual masters who neglect their bodies while developing spiritual powers.

4. Measures of Science and Vigyan

Progress in science is measurable by scientific discoveries, inventions, and Nobel Prizes in sciences. Science is collaborative. However, ***yog vigyan*** is absolutely individualistic. In particular, can any two yogis optimize their spiritual powers? This is an open question. In recent years, Maharishi Mahesh (1918-2008), the founder of Transcendental Meditation, had carried out experiments in a golden dome erected on the campus of Maharishi University of Management, Fairfield, Iowa. His crisp formula is that the square root of one percent of the population of a community can bring 'harmony' around it. Buddha, who lived in a period 563 BC - 480 BC, was known to create such vibes in places wherever he camped during his discourse circuits.

5. Shadows of Science and *Vigyan*

In this context, it is easy to see the impact of science, say, of the US on India especially in terms of the US iphones, entertainment and medicine. On the other hand, the ***yogic vigyan*** of India is a big-time holistic health symbol in the US. Since the 1960s, yoga has become an industry in the US - yoga pants, toga leotards, yoga mats, yoga sandals and yoga tops have become a public rage.

Bites and Bytes

Participation in science is like being a part of a huge musical symphony of scientists. Modern science is not solely confined inside the four walls of a school, college, or university, but it thrives in the garages, barns and backyards.

Finally, a question arises as to the existence of any attempt to synthesize scientific power and yogic power. To the best of my knowledge, the Nazis alone, under Adolf Hitler, have tried it. Its evidence was seen in the presence of many yogis who were invited from India. They lived in the underground bunkers in Berlin during World War II. The sole objective was to 'harness and optimize' the yogic power to influence the course of war. They were discovered after the allied forces took over Berlin in April 1945. I have not seen any report on their mission. Anyway, the Nazi Swastika, a slight modification of the Swastika of Hindu religion, is a symbol of this synthesis too!

June 14, 2018 (Chandigarh)/March, 2021

48. SCIENCE AND *VIGYAN* ANALYZED

[**Note**: In an article, ***SHADOWS OF SCIENCE AND VIGYAN***, written three weeks ago, I differentiated between science, as identified since the Newtonian era, and ***vigyan***, as traditionally understood in India. Scientific thinking is epitomized in the inference of identical conclusions when an experiment is performed under identical conditions. This article was never on my intellectual radar, but due to recent insights gained in the ***Vipassana*** meditation course, this article became a natural sequel.]

In the history of ***Vigyan***, as far as I have understood it, so far, there is no evidence of the existence of instructional laboratories in ancient 'schools' and 'colleges' during its initial phase. Instead, the entire emphasis has been on the cultivation and harnessing of the ***Praan Shakti,*** called the **Life Force.** The Patanajali's ***yoga sutras*** (196 of them) have spelled out a course for realizing the yogic powers at the individual level. It goes back to the period around 400 AD - during the Golden period of India under the Hindu kings. It should be understood that the treatise of ***yoga sutras*** was the culmination of ongoing efforts. However, they crystallized under the name of Patanjali - as an individual, his academy, and his brand name too. It is like the works of the ancient Greeks under the names of Pythagoras and Euclid.

A point that needs to be stressed is that once the Hindus lost their homeland politically, yoga became a victim of stagnation, superstition, and distortion. My seminal effort is to understand ***Vigyan*** in the context of yoga and meditation as experienced during the 10+2 days (June 20 - July 01, 2018) of the ***Vipassana*** course. In fact, I repeated this course, as I had done it first in September 2007. Here are the nuts and bolts of ***Vipassana*** meditation in the context of scientific experiments - the numbering is permutable:

1. In the study of science, say at the college level, a laboratory for 20-30 students is different from a typical classroom. Boys and girls work together. In the Vipassana meditation hall (at Dehradun), nearly 70 meditators sit on identical cushions - some needing back support too. Men and women

sit in the two halves of the hall facing the assistant teachers. The two environments are totally different.

2. The segregation of gender is practiced to an extreme - that even at the time of dining and strolling, men and women can't see each other. However, the ***Vigyan*** of meditation has deeper reservations on gender and sex. On the other hand, science is gender free, though its pursuit may not be even. By the way, an answer to a perennial question on sex lies between the extremes of suppression of sex - sublimation of sex, or immersion into sex.

3. In physics, chemistry or biology, an experiment is clearly stated. There is no confusion about it. However, it is not the same in the ***Vigyan*** of Vipassana meditation. For instance, the second technique, as explained on the second day, '***observe sensations, as they rise up and fall in the triangular area formed by the nose and mustache area above the upper lip***' is perplexing, though each meditator begins to get its feel after ten hours of practice in one day.

4. In a science lab, students are issued identical equipment, chemicals etc. and they work on their experiments quite closely. The lab assistant or demonstrator is there to assist, if a student is not able to follow certain directions. However, it is not the same in meditation. For instance, in the above meditation technique, the body sensations are infinitely many, and they are different in different persons. Furthermore, even in the case of one individual, they are unpredictable and may vary every minute.

5. Duration of an experiment is a significant factor too. In a typical college-level experiment, its duration is three hours - including making observations, doing some calculations, and writing inferences drawn. In contrast, the above meditation technique was practiced the very next day for 10 hours with a break of 5-7 minutes after every session of 55 minutes.

6. A science lab assistant or demonstrator openly resolves students' questions for the duration of the experiment. In Vipassana meditation, for a 10-hour meditation practice of one technique, two 30-minute time slots are allotted on a one-to-one basis. No judgment call is implied here. At the age of 78, I have figured out that the ultimate resolution of

a question has to come from within. This is more true in the domain of meditation.

7. Science does not put any restrictions on its seekers in terms of food and drinks. That is not the case for any meditation. In Vipassana, the food is all vegetarian - even the use of onion and garlic are excluded!

8. In science, there is no room for any 'social' un/non-touchability. Of course, scientists are watchful for any external contamination getting into their experiments. During Vipassana meditation, no one was supposed to touch any other - not even casual greetings. Furthermore, everyone used the same set of steel utensils - a plate, two bowls, a spoon and a glass for all the 12 days. You clean them up yourself after every meal and place them in an allotted space. The idea behind this ritual is to enhance the overall sensitivity of the body and mind.

9. This one is a corollary of the 8th above. It is believed that each person has a unique signature vibration, and one may lose it through physical contacts. In parapsychology, there is an aura of a person. On the contrary, science and mathematics thrive in close collaboration. Yes, all human organs are in perpetual motion, yet their cumulative impact in the surroundings is very subjective and debatable.

10. Collaboration between scientists and mathematicians in the offices, labs and bars is proverbial, but nothing like this has a place in the Vigyan of meditation. In contrast, in Vipassana meditation, Arya silence is strictly enforced and strongly recommended. Arya silence is essentially silencing the body completely from within and without - no talking, no reading, no writing, no sign language etc. While walking alone, just keep your gaze on the grounds.

11. Given a human problem - be it that of health, transportation, or construction, scientists and engineers pool their expertise and come up with a solution. In the Vigyan of meditation, however, it is an open question - as to how to apply the yogic power individually or collectively to discover or invent something new. Most Hindu scholars aggressively point out a nebulous Vedic mantra(s) whenever a breakthrough discovery makes headlines in the media. But no one has ever dug out a principle of

science or a theorem in mathematics from the Vedas that has never been known before.

12. Those who take a science course(s) exhibit their proficiency according to their background and labor put into the learning of the material. Students in science and math courses are routinely graded at the end. But this is not the case in the Vigyan of meditation, say in Vipassana. It is nearly impossible to grade the meditators after 10+2 days. Part of the reason is that the prerequisites for mediation are subjective. Thus, at the end, any two individuals in the ***vigyan*** of meditation may end up poles apart in their outlooks on life.

In conclusion, science and Vigyan are different and far apart from each other. I have enjoyed digging around this topic for three weeks. I hope scholars on both sides pitch in for a healthy exchange of ideas. Nevertheless, even the superstitious Hindus from India have won the highest awards in science when they move to the West - say in the US or UK. However, this is a different topic altogether.

July 08, 2018 (Bathinda)

49. HISTORY OF MATH & HINDU RENAISSANCE

"India Spices, a section in the Darts, has eight reflections - mainly dealing with mathematics in ancient India, colonized India, and contemporary India. Give a critical review of these reflections and identify specific features that stand out in your mind." Darts, in the first line above, is the abbreviated title of my book, ***Darts on History of Mathematics*** (2014), which is used as a supplementary book in Math 314, an History of Mathematics (HoM) course that I am currently teaching. This question was for an in-class write-up to be finished in 10 minutes. Such write-ups in an HoM course correspond to the quiz problem(s) that are given out in typical undergraduate math courses.

The reading of varied responses of the students was amusing at times, surprising at others, and on a couple of occasions, they stopped me to think. The main reason for the wide range of responses is that India is not more than a geographical entity for most Americans. Furthermore, the US students are getting increasingly centered in their digital worlds, where India may flag into their short attention span fleetingly.

However, this reflection is prompted by the following verbatim comments from the write-ups of two students: "***It was great that you brought your own culture into terms with the HoM***" and "***The section of Indian spices is that to you, but not just you. If you are like me, and view all ancient history as our collective shared past, then it is insightful in regard to ancient India.***" I am not trying to analyze these comments or draw much of a meaning out of them. At the very instant of my grading these responses, there was a surge of the following thoughts, which may have been sitting on the edge of my subconscious mind - call it a trigger effect.

I said to myself that if I won't raise the awareness of the public about the mathematical contributions of ancient India to the world, then who else is more qualified to do it? Also, Gandhi's popular saying came up in my mind - you become the instrument of change that you want to see in the world. That is what I have since taken upon myself.

In all, the ***Darts*** have seventy reflections which are unevenly divided into five broad sections. There is only one reflection, **Mathematics in the Vedas** that I had written up in 1988. It came out of my teaching a short course, **Fast Mathematics, during** a three-week mini term held in January 1983 (now discontinued). My lectures were based on the book, ***Vedic Mathematics*** (1965) by Bharatiksna Tirath (1884-1960), who was the Shankaracharya of the Goverdhan ***Mutth (***Puri*)*, one of the four ancient seats of Hindu philosophical and religious thoughts. Before joining the order of Shankaracharya, Tirath had also done a master's in mathematics. During my sabbatical leave (1986-87), I did extensive research on Vedic Mathematics - mainly through discussions with several scholars living in different parts of India. The reading of the Sanskrit texts was out of question for me as being out of India since 1968, I never got to study the Sanskrit language in any systematic way.

At my age nearing 80, one thing that I can say loudly and clearly is that the Vedas provide a template of multi-dimensional intellect, which alone can extract various meanings embedded in different layers of the Vedic mantras composed in verses. The tragic irony is the continuous ostracization of Sanskrit in the school and college curricula in India even after 70 years of India's freedom from the colonial yoke. The present Indians are not ready unless a drastic change is brought about in India's education system, similar to the radical changes in the policies of Demonization (2016) and the Goods and Services Tax (2017) that have overhauled the monetary system of India. It was mainly due to the vision, sagacity, integrity and leadership of Narendra Modi, the Prime Minister of India (since 2014 -). It is equally an electoral gamble too.

In the spirit of the famous poem, ***Let My Country Awake***, by Rabindranath Tagore, the first non-European Nobel Laureate in Literature (1913), about the awakening of India in freedom, I would say that only the multi-dimensional intellectuals will herald the awakening of the Hindu masses. The Hindus have made up 80-100% of India's population during the last millennium, but they have lived under various forms of foreign subjugation in their own homeland! Free thinking amongst the masses lags far behind their political freedom.

That is how my mind was ignited by the aforesaid comments which may even appear wayside to some readers. In my experience, HoM fully comes to life when it is approached as a subset of social studies.

March 08, 2019

COMMENTS

Satish Ji, Your article is provocative and very much to the point about the Hindu culture - strong and weak points. Just as lofty as it goes, the Hindu concept of ***'Ishta Devata'*** (from Patanjali's Yoga) gives freedom of belief. since the word 'belief' means just that, that it is not something scientifically proven. However such beliefs offer comfort and guidance to devotees until a better concept takes over the prior. However we do not use the 'belief concept' when we practice deep breathing or meditation or do yoga asanas. We do not say that 'we believe in breathing' since it is self - evident.

Only in Hinduism the concept of freedom of belief is a part of its culture. Every Hindu household can have a multitude of gods and goddesses. It is unfortunate that Hindus have extended the same rationale to the 'Monotheists - Christians and to Muslims without realizing that any deviation from their institutionalized belief in Jesus or in Mohammad invokes in them their Biblical or the Quranic concept of sin and 'sinners' for those who do not believe just as their holy books command them. The same command urges their followers the virtues of killing the 'sinners' to earn heavenly rewards. The Portuguese or the hordes of Muslim invaders from across Khybar were singularly united to kill the 'non-believers' to earn heavenly rewards. Hindu's rational defenses were no match to the monotheist'š fanatical beliefs! The sad part is that to this day in the 21st century there are hundreds of pockets of 'fanatical believers' carrying out the violence believing in the same hateful 'Quranic' concepts.

Coming down to the issue of Math and India, is it true that the Hindu scholars developed the 'Pythagorus' geometric equations? Also India is known to have developed the concepts of numerals including 'zero' and decimal. With my best wishes. **Vijay Kapur**

You are doing a great job at sharing in an interesting manner. Thank you. **Brian Winkel**

Dear Dr. Bhatnagar, Thank you for sharing. Slowly I am trying to include more culture in my instruction. My students are doing a project on Ethnomathematics and they are enthused to explore how other cultures valued mathematics. Thank you, **Renuka Prakash**.

50. MADE IN MATHEMATICS!

A week ago, while chatting with the Chairperson of the Mathematics Department of Panjab University (PU), Chandigarh, my Alma Mater (MA, 1961), I learnt that the student intake (2019-20) in the Master's Degree Program was 70 (not including Statistics). This number is determined by some formula that I did not care to delve into. However, I did recall that the class intake was 25 when I was admitted in the 1959-60 session. For a broader perspective, the number of 'math loving' students, in the then undivided state of Punjab - comprising the present states of Haryana, Himachal Pradesh and Punjab, doing master's from the affiliated colleges of PU - in Amritsar, Jalandhar, Ludhiana and Patiala was not more than 100.

I told the Chairperson that the number of students doing master's in mathematics and statistics in the two universities (UNLV and UNR) of the state of Nevada is hardly 30, and on average, about a dozen of them finish their degrees each year. For a comparative examination, the population of Nevada is 3 million (in 111,000 sq. miles); whereas, the population of Chandigarh Capital Region is 2 million (in 500 sq. miles). Here is a shocker - the number of students doing their master's from the colleges and institutes within a radius of 10 miles of the PU is nearly 300! An interesting demographic feature of these students is that at least 75% of them are female!

One has to pause here to understand what is going on with respect to mathematics in the US, technologically speaking, the most advanced country in the world today, and in the emerging India, which got its freedom from the British, its last rulers 70 years ago. So many questions pop up in my mind. I do have a unique vantage point for having lived long enough in these two cultures.

What kinds of jobs are waiting for these graduates in India? It appears dismal. Smaller government, a perennial stand of the Americans, is taking hold in India too - implying fewer government openings. Job opportunities in the young private sector have a long way to go, particularly in research and development. In the US, where I live, all the math graduates are absorbed in high schools and colleges. A few of them do go for PhDs in Math Education and other areas of Education, rather than going for traditional PhDs in mathematics and statistics.

The deeper as well as the wider question is the need, value or worth of higher education in general. The US and India are both democratic nations. The US is capitalistic and perhaps, the only such nation in the world, but India has been very socialistic in the beginning. Consequently, higher education in the US is pursued by the ones whose heart and soul are into it. India is the only country in the world where education, from nursery to tertiary, has become a mighty industry, an engine of the national economy. That has been creating a class of constipated and unemployed "educatees". Lately, this class has been politicized as they have started demanding jobs or unemployment compensation! Ironically, Indian society in general and the Government Labor Department do not fully recognize the individuals plying mom-pop types of trades, though they do generate small jobs for others as well as for themselves.

Coming back to the production of MS/MA (Math) which India seems to be producing far more than the US by a factor of ten, a question arises as to how far it carries over into the caliber of fundamental and seminal research in mathematics done by the Indian mathematicians in Indian universities, as measured by the winners of mathematics awards like, the Fields Medals and international ranking of mathematics departments etc.

Well, raising a question is often easy. In this scenario, I would venture to add that a steady state in degree production will be reached in higher education in India. It would require political will, free enterprise and full autonomy of academic institutions. Again, this is just the tip of the iceberg of the monolithic higher education in India.

May 31, 2019 (Ahmedabad)

51. ETHNOMATHEMATICS OF INDIA

[**Note**: This article has evolved during the last six months as it underwent different levels of thought processes. Started in August - on noticing the announcement of a session on Ethnomathematics at the annual Joint Mathematics Meeting (JMM), to submitting an abstract of a paper, '**Ethnomathematics of India**' in September, and to its acceptance in October. Since then, the topic broadly raced through my mind until it was not presented on January 18, 2019. JMM is billed as the largest show of mathematicians in the world for drawing nearly 6000 persons. For the purpose of its inclusion in this volume, it was essential to clean up the paper and expand it, which was not possible at the meeting due to limited time.]

Understanding Ethnomathematics

On the face of it, Ethnomathematics is formed out of the two words, 'Ethnicity' and 'Mathematics'. It simply means mathematics associated with a culture or a group. Depending upon the geographical area, it may also apply to a small nation or even a civilization. According to an online source, the term "Ethnomathematics" was first introduced by Ubiratàn D'Ambrosio (1932-), a Brazilian educator and mathematician in a 1977 meeting of the American Association of the Advancement of Science.

Ethnomathematics is one of the extremes of Eurocentrism in the history of mathematics. For instance, George Varughese's book, **The Crest of a Peacock,** traces the roots of modern mathematics to India long before Newton. Ethnomathematics is also viewed as a mathematical spin-off of diversity in conversations in every aspect of the present US society.

Ethnomathematics is a body of mathematical skills, tools, and practices which are absolutely unique to an indigenous community, society, or even to a nation. It is mathematics which is practiced among identifiable cultural groups such as national-tribe societies, labour groups; the study of mathematical ideas of a non-literate culture; the investigation of the traditions, practices and mathematical concepts of a subordinated social group.

There are two-fold objectives of the teaching and researching in Ethnomathematics -contributing both to the understanding of a culture and to the understanding of mathematics embedded in it - leading to an appreciation of the connections between the two. Thus, a humanistic side factors into the study of Ethnomathematics

State of Ethnomathematics

In a way, **Ethnomathematics** is similar to **multiculturalism**. But there is a big difference, whereas multiculturalism, in the last fifty years, has become a part of mainstream USA - in the academe - from curriculum to administration to political correctness. However, Ethnomathematis has not gotten any traction or has made any waves.

It is worth adding that **Ethnomathematics** is different, say, from Chinese Mathematics, which is understood to be a body of mathematics that is historically developed by the Chinese mathematicians. Likewise, say, the Jewish Mathematics, which was first developed by persons of Jewish religion. However, mathematics remains universal, as it is always independent of national origin, religion or geography.

History of Ethnomathematics

Being interested in history in general, and its subset, history of mathematics, it is natural to pose this question. However, an answer is simple, if we assume that the rudiments of mathematics - like counting and understanding of some geometric properties like that of circles, right angles and right-angled triangles are innate to human intelligence, then its history can be traced back to any time in the past and in any place or culture. There is something cognitive about properties of natural numbers, and axioms of Euclidean geometry.

Ethnomathematics of India

Vedic Mathematics is the Ethnomathematics of India. The word Vedic is derived from the Veda, a Sanskrit word, also means knowledge. Interestingly, Veda is used for one Veda as well as for all the four Vedas, namely Rig Veda containing 10,552 mantras, Yajur Veda containing 1975, Saam Veda containing 1875, and Atharva Veda 5977. In simple terms, a mantra is a set of energized syllables.

The language of the Vedas is Sanskrit, which is different from the prevalent Sanskrit. It is rightly called Vedic Sanskrit. The power of Vedic Sanskrit is that different meanings can be layered or embedded in one mantra. Thus, mathematics is found as one the embedded meanings in the mantras. The Vedas are the holiest scriptures of the Hindus - coming down from ancient times. The dating of the four Vedas remains debatable.

For a number of historical reasons, the Hindu culture became dormant due to hundreds of invaders and marauders from Central Asia, Middle East and Western Europe. Subsequently, Hindus lost all their treasures- from their grandiose temples, to institutions, to all ties with their heritage. In particular, Vedic Mathematics disappeared from the cultural horizon of India for the last one millennium.

In 1964, the Vedic Mathematics got international recognition, when Bharati Krishna Tirath delivered a series of lectures on Vedic Mathematics in the US. In 1965, he published the book, **Vedic Mathematics**, which has been sold in millions. Essentially, Vedic Mathematics is a set of eighteen ***sutras***, aphorisms, for solving arithmetics problems.

Tirath claimed to have extracted these ***sutras*** from the appendix of Atharva Veda. However, to this day, no one has been able to trace the exact origin of these ***sutras*** to any one of the Vedas. It is an open research problem in the realm Vedic Sanskrit, linguistics and mathematics.

Finally, in the 1980s, Vedic Mathematics got a boost from Maharishi Mahesh yogi of the Transcendental Meditation fame, who supported its propagation. In the late 1990s, the Government of India, under the NDA coalition, tried to implement its introduction in schools.

June 02, 2019

SECTION IV

SMORGASBORD BITES

52. SCIENCE & MATH IN STELAE

"Get a few good shots of the images that can be used in my book on the History of Science that I am about to finish", thus, said Alok Kumar, a physics professor in the New York University at Oswego. Twice, we talked before my departure for a one-week archeological study tour of the Mayan hieroglyphs and iconography in Central America. Alok was more confident about my ability to find some images than I was about myself. I am not even an amateur photographer. Lately, I started taking pictures especially of the 'muscular' trees - dead or alive, as they project the overall robusticity of the people who get my admiration. So far, it does not apply to archeological ruins - no matter how grand they might have looked in their glorious past!

Alok's interest in History of Science runs parallel to my interest in the History of Mathematics. Modern science is rightly written in mathematics. Galileo (1564 - 1642) was the first scientist to assert it at the dawn of modern science. Science and Mathematics being inseparable means that any archeological finding of one would lead to the discovery of the other to some extent. Such was my mindset before and during this trip.

Another line of thinking for the trip was that if science is studied quantitatively, then math has to be there in a gross form, as seen presently even in the case of economics when it is studied quantitatively vs. qualitatively. In humanities, math is hidden under thin layers. The Vedic science and mathematics is a perfect example, and so it is in many ancient civilizations like Egyptian, Chinese, Inca and Maya. I am not a scholar of any one of them, but have spent nearly 30 years reading about them.

Stelae (***shilalekh*** in Hindi - means inscription on stones) immediately reminded me of nearly100 of them that the great Indian Emperor Ashoka (304 BC- 268 BC- 232 BC) had them especially commissioned and strategically installed on major highways in his vast empire covering almost the present Indian subcontinent. During the trip, I saw some of the nearly 500 Mayan stelae that have been discovered since the middle of the 19th century when the Mayans made the world headlines. They are displayed inside the museums and also placed in the open archeological sites in Copan and Quiriguan in Guatemala. A couple of stelae were precisely rectangular - a block of granite, fully covered with glyphs. They have preserved some historical events and knowledge of the medieval era of the region.

The excitement of Michael Grofe, our guide (who is also an archeology college professor) on seeing these stelae and glyphs would fly off to other participants, but my mind mulled and hovered over the following points and conjectures:

From the stelae, ruined palaces and pyramids, it is fairly evident that the Mayans had a deep knowledge of geochemistry, geology and sedimentary rocks etc. Certainly, modern super specialized categories may not be there then. For instance, for stones needed for grand construction projects, the choice of a mountain, identification of queries in it, selection of the right granite, its mining in humongous blocks (some in a couple of tones) and their transportation - require a lot of engineering. Cutting of the rocks in precision is an enigma to me, and no less are the tools of cutting through rocks. In some places the distances of the mining sites from the parking places were a couple of miles. Egyptian donkeys cannot haul them, Thai elephants are not a part of the forest habitat of the region, and the argument of slave labor is ridiculous.

While walking amidst archaelogial alleys, I hummed the tune of a popular Hindi movie, "***Geet gaya patheron ne***" means that 'stones are singing'. It is the theme song of the movie too (1964). I strongly believe in the presence of mathematical DNA in intricate stone monuments. If scientists can extract molecules of water embedded in the rocks so math historians can extricate mathematics out of intricate designs and architecture of stone superstructures. However, such a scholar must be an all rounder in science, mathematics and history or a team of dedicated specialists with interdisciplinary passion for discovery.

August, 2012

53. PROPOSAL - HoM SESSION AT THE AMS MEETING

(A Note to the Session Coordinator)

Thanks for communicating before the deadline. Let me tell you that the moment I saw the requirement of potential speakers, I had no idea of anybody else except me! Reason: over the years, I have organized 3-4 MAA contributed paper sessions singly and jointly. The merits of a proposed session was the sole criteria. They attracted 10-20 participants without my knowledge of any one of them. The AMS must have some rationale for it - but that is beyond me. My understanding is that neither potential speakers are paid, nor their papers accepted automatically.

Since it is a session on History of Mathematics (HoM), let me add a bit of its history, if you don't mind knowing my version. First of all, your handling this business over the weekend is appreciated. To the best of my knowledge, through Y-2012, no US university has a department of History of Mathematics. This was a finding of a hands-on history project that I required while teaching UNLV's graduate course, ***History of Mathematics*** (MAT 714).

Its corollaries are that faculty are never hired in HoM as their specialization. HoM is not a concentration in a graduate degree program, except a very strange one in Virginia. Despite these odds against HoM, at every JMM, the sessions on HoM are organized by all major mathematics organizations - including AMS, MAA, and SIAM. The number of participants and the papers presented may be more than in all the traditional disciplines combined! It is an open research question to explain this popularity of HoM.

Neither, I am known in a mathematics circuit, nor get invited to give talks. When I joined UNLV in 1974, it was primarily an undergraduate institution; its master's program in math was only 3-4 years old. Today, the math PhD program is already 9 years old. Call it my personal or institutional isolation, knowing for me interdisciplinary scholars on the proposed theme is not easy. Yet, seven names have cropped up eventually. If you know someone seriously interested, then it may be added too.

Generally, sessions on HoM are focused on Europe and time lines ranging from the 16th to 20th century. An emeritus math professor and founder of two journals just emailed (without knowing my proposal): ***I suggest you organize a session based on this as it is richer than the usual history of math talks***. This theme intends to cover the entire globe for the presence and absence of mathematics. The time line goes back to the BC era in Greece, India and China. Accepting this proposal will spot light HoM in a desert! Finally, diversity has to be supported at every level.

The attachment has a page from my forthcoming book, ***Darts on History of Mathematics***. It has some tangential connections with the proposal. That is all.

September, 2014

54. A HUMANISTIC MATH REVIEW

Four months ago, I went to UNLV's Lied library in search of a new textbook for an undergraduate course, History of Mathematics (Math 314), which I am scheduled to teach in Spring-2015. At the same time, I pulled out a couple of books from the shelves for my supplementary readings. After browsing the prefaces and cursory glances at some pages, I returned all the books except, the ***Ways of Thought of Great Mathematicians*** by Herbert Meschkowski (1909-1990). It was published in German in 1961, and translated into English in 1964. That was a kind of plus factor for getting my attention. Moreover, since its acquisition by the library, from the circulation stamps, it was interesting to note that it had been checked out 50+ times – a readership rarity.

The title of the book suggested as if the author had succeeded in understanding the minds of the nine mathematicians included in his study- from Greek antiquity to the end of the 19th century. A person cannot understand one's own mind, so how can he/she understand those of the others? Still, my curiosity about the book stayed. The following few lines in the Preface got me going: "***To understand the fundamental problem of modern mathematics, one must study the history of mathematics..... the modern mathematics lives in a sort of ghetto, in voluntary isolation, out of reach problems which cannot be attacked by mathematical methods***."

Initially, I thought the author was an eminent research mathematician, but he turned out to be a multi-dimensional educationist, who taught in high schools and universities, authored many books, and served as president of a university. His name got confused with the famous German research mathematician Hermann Minkowski (1864-1909). Talking of Germans, Besides Pythagoras and Archimedes, six out of seven mathematicians are Germans. The French child prodigy Blaise Pascal (1623-1662) is an exception. It may speak of the author's bias for German intellectual superiority.

The nine lives of featured mathematicians take up only 100 pages. For me, there was nothing to get interested in their mathematics that is sketched out selectively. I look into the total environment that prompts a particular kind of research problem(s). Generally, my focus in the history of math is to find the necessary and sufficient conditions in the making of a great mathematical mind and institutional culture. That I call 'theorems' in the history of mathematics.

Despite having read a lot about so many mathematicians, I noticed at least one refreshing nugget in each write-up. Either, I had not heard of or completely forgotten his choice of Nicholas of Cusa (1401-1464), both a cardinal and mathematician. His thoughts resonated with mine when he said, "***If we can approach the Divine only through symbols, then it is most suitable that we use mathematical symbols, for these have an indestructible certainty.***"

From the ancients, his choice of Pythagoras and Archimedes, and his lauding of the Greeks for the development of axiomatic mathematics, it all comes out of a conditioned Euro-centric mind. To me, it shows a general ignorance of the achievements of mainly the Chinese and Indians who lost their heritage due to European colonization. Overall, the book stimulated my thoughts, and, at my age of 75, that is a good measure of liking a book. The book is within the reach of all college students.

January 03, 2015

COMMENTS

Satish, Interesting, although I never really get excited about history of mathematics per se, I rather think that for my students discovering old principles as their own while oblivious to history and then finding out that some bigshot did the same thing, e.g., discovering Fourier Series through least squares approach on their own - they do it every time. And as for quoting from Catholic cardinals, like Nick AKA Nicholas of Cusa I would not go there, considering their recent track records on truth!!! Take care and keep writing!!! **Brian**

Aadaraneeya Prof. Bhatnagar: ***Jaya Siya Raam***. Thanks for your mail. I am happy to see that you are never out of passion and interest in reading fascinating material. I am lured to read that book. I hope you are keeping well. With best wishes for a happy & prosperous new year. **SL Singh**

I'm not likely to read the Meschkowsky book, but I've read about 30% of a book that could serve as a text for a history course meant to encompass science, math, and world events in general. I doubt that you would seriously consider it for a primary text, but it certainly merits listing in "related readings." The book is ***A Short History of Nearly Everything***, by Bill Bryson. It was published in hardcover in 2003, and in paperback in 2004. The copy I have is paperback, and from the 11th printing, if I understand the notation correctly. It's worth a bit of your time, and you'll get a quick idea of it by starting with Part II, entitled "The Measure of Things." Best wishes...Happy New Year, **Owen Nelson**

55. EXPANDING HoM – HISTORY (PART I)

There is a trilogy of ***mathematical reflections*** which has been occupying my mind for the last 3-4 months. It seldom happens in my writing tempo. Its reason is simple. The moment an idea of a new reflection ignites my mind it immediately goes into the highest gear and locks it up with such intensity that the previous reflections become stranger to me within a few hours - in the sense that I cannot even recall them out rightly. However, this trilogy is so stubborn that its deliverance alone will 'empty' out the womb of my mind!

For the last ten years, intellectually as well as pedagogically, I have been flying high on the history of mathematics (HoM) – whether while teaching a course or contemplating on it. The reason being, I view HoM as a subset of history in general. As an academic discipline, History is the most social one - both vertically and horizontally. Apart from my ongoing readings and writings, a certain maturity that comes with age (now 75+) has kicked into my thinking too. This enhanced social awareness is expanding my mathematical thinking into far-fetched disciplines.

To the best of my knowledge, no one has approached HoM in this manner. For example, traditional textbooks on HoM and science are totally devoid of it – irrespective of whether written by academicians in mathematics or in history. Often, their intersection is close to an empty set. Of course, there is bias when they deal with HoM of their erstwhile European colonies. I see it all the time whenever a European writes about HoM of India.

As a result, during the last four years, I diligently worked on the publication of my two books - one purely on history and the other specifically on HoM in order to establish my academic credentials as an historian as well as to aver a new thesis in the teaching of HoM courses. About two years ago, I went public with my ideas at a regional meeting of the American Mathematical Society held at the UNLV campus in April 2015. I proposed, organized and chaired the first ever session on HoM with a new theme.

The papers were solicited on the necessary and sufficient conditions for the development, growth and flourishing of mathematics in an individual, department, institution, society, or a nation – both in historical and in contemporary settings. It has set the ball rolling further that I have started thinking of proposing a graduate certificate program in HoM. It will include one graduate course from each of the UNLV's departments of History, Political Science and Philosophy. Such a program seems to be non-existent in the US. At least, my Google search did not produce anything in this regard. This will definitely enhance the depth and breadth of math and science teachers in high schools and four-year colleges.

During the Fall-2105, I offered ***History of Mathematics*** (MATH 714) for the fourth time. I have always invited guest lecturers from the History Department in order to provide a different perspective on HoM. In this regard, the co-operation of the History faculty is highly appreciated. Because of the way curricula are set up, History majors rarely take a mathematics course beyond pre-calculus. Nevertheless, over the years, Professors Bubb, Moehring and Whitney have lectured in my HoM courses and Honors seminars. Invariably, students enjoy new ideas, presentations and points of view from a guest speaker who is often distant from mathematics. Let me add that HoM course is not a regular mathematics course requiring any specific mathematics course as a prerequisite.

For a certificate program in HoM, I am looking for one or two regularly offered graduate courses in history, which would highlight the golden periods in science and mathematics, especially in a particular epoch, nation or civilization. In a way, this is a tall order. Generally, the expertise of faculty in each department clusters around 4-5 research areas, as it facilitates the running of healthy graduate programs requiring theses and dissertations besides the course work. There has to be at least 2-3 faculty members in one research area.

Great empires have flourished in Africa, Asia, South America and the Middle East. I have no idea of Australia and New Zealand before the 18th century, when European migration started there. By and large, modern academic foci in HoM are in Europe and North America. Nevertheless, a couple of good history courses should not be hard to identify for the certificate program.

February 25, 2016

COMMENTS

Dear Satish, Thanks for your email. It's always great to hear from you. I'm copying Paul Werth, who is now Chair of the History Department. He might have some ideas about how to pursue a potential partnership. All my best, **David Tanenhaus**

56. IS MATHEMATICS SELF-TAUGHT?

One of the problems, included in the Midterm Take Home Exam in MAT 711 (Survey of Mathematical Problems), emerged when my students and I watched two videos on Srinivasa Ramanujan (1887-1920). ***The Genius of Srinivasa Ramanujan*** (2011) is a 60- minute documentary and ***The Man Who Knew Infinity*** (2016) is a 108-minute regular theater movie. The documentary sets are largely Indian, but most movie shots are done overseas. Both are excellent artistic productions.

From a perspective of this course, the fundamental question is: is it possible to learn mathematics to a level at which Ramanujan did without anybody's help and guidance? Let me state at the outset that in the world of human beings, it is very easy to raise questions than find answers. In the course textbook, ***What is Mathematics?***, the author, Richard Courant has written four pages on **What is Mathematics?** before the opening of Chapter 1. Moreover, on Day #1, students were encouraged to write a brief in-class essay on **What is Mathematics?**. Thus, the ground was fertile for the exploration of this question.

There are instances in every culture and society, where precocious kids have shown mastery of natural numbers and their properties. Adults are known to teach themselves arithmetics and its applications - especially in the seclusion of homes or prisons. It speaks highly of both the intuitive and deductive nature of mathematics which may fully interface with some individuals, whose brains are naturally 'programmed' in that orientation.

Over the last forty some years, I have heard of Ramanujan's mathematics - how it has taken five decades for world class mathematics to figure out the proofs of the theorems that Ramanujan had simply written them up in his notebooks without giving any proofs. But it was not that simple. Paper being very scarce and expensive in India of the early 20th century, Ramanujan first worked his math on a slate with chalk.At the end, he neatly copied the results in his notebooks. On a personal note, we hardly used any paper during my elementary school (1945-50) in Bathinda.

Of course, every time Ramanujan was asked this question, as to how he had proved a particular mathematical result, his simple answer was that his family goddess, Namagiri had revealed it to him. We are all conditioned by our surroundings. Tamil Nadu is India's southernmost state- known as the land of purest form of Hindu religion. Harnessing the mental power through various techniques of Yoga in the pursuit of any discipline is known to have produced baffling results. Ramanujan harnessed his power through ***Bhakti*** yoga, the devotional path.

One thing is clear that it was only after his association with Hardy and Littlewood (1914 -1919) that Ramanujan understood and finally grasped the important concept of proof in mathematics, otherwise, no mathematics journal would have published his results. The point is that had Ramanujan lived long enough, his way of discovering mathematics might have been unraveled. It is a psychological conjecture!

Finally, where do I stand in the understanding of Ramanujan type of persons' ability to do mathematics on their own? My guess is that their brain cells are wired in a very unusual configuration. It is suggested by a Yahoo news item of the day. The story is of a boy speaking fluently in a new language after an injury to his head. He forgot the language that he knew before injury, and woke up talking in a different language. Back in north India of the 1950s, I had heard of a man, not knowing Sanskrit, but he gave discourses in Sanskrit language in a state of trance. The great American clairvoyant Edgar Cayce (1887 - 1945) solved all kinds of human problems by going into a self-induced hypnotic state. The following quote sums up the state in the field of foreign language syndrome: ***"It's an impairment of motor control," Dr. Karen Croot, one of the few experts in foreign accent syndrome, told CNN a few years ago. "Speech is one of the most complicated things we do, and there are a lot of brain centers involved in coordinating a lot of moving parts. If one or more of them are damaged, that can affect the timing, melody and tension of their speech."***

Self-learning of music is understandable and is seen commonly. However, the making of musicians like Mozart is not easy to figure out. Mathematics will continue to defy the experts for a long time. But these 'experts' must be knowledgeable in mathematics to some extent, which is counter to the disciplines like psychology and neurologists. That is where I rest my pen as far as the answer to the take-home question is concerned.

October 24, 2016

57. AN HOLISTIC BOOK REVIEW

ASTRONOMY AND CALENDARS -THE OTHER CHINESE MATHEMATICS

Authored by Jean-Claude Martzloff. Springer - Verlag (2016)

[Note: The Mathematical Association of America has an active book review program. Considering it as a professional obligation, I chose to review this book out of a list of half a dozen mathematical books that was given to me.]

As an author myself, the first thing that I look out for in a book is its preface. At my age of 75+, I do not want to waste my time randomly on a book - whether for reviewing it, or reading it. This book has no Preface! Out of a total of 472 pages in hardcover, 134 pages include seven Appendices, Tables of Chinese Calendars, Primary and Secondary Sources, Glossary, and Index. For researchers in Chinese Calendrical studies and astronomy, this book is likely to be useful. But, I am not aware of a US college course, where this can be used as a textbook. Sure, it is the author's labor of love.

This book was first published in French in 2009, and this English version came out in 2016. Strangely, it is not explicit, if the two authors are the same! Why not have this thing cleared up front? It is said that an author's clarity of ideas is lost in the least only in the translation of a mathematical work. But in this book, I wrestled with the meanings in several passages. The translated English seems like Frenglish, like the **Hinglish**, a hybrid of Hindi and English languages. Hopefully, in a revised English edition, the translation may be improved.

In the Acknowledgements, supposedly, the author (no name, no date or no place at the end; the three coordinates of any write-up) states that the writing of this book was undertaken, as there was no one-stop book on the Chinese calendars and Chinese history of mathematics. It reminded me of how in the European- colonized world, all the books in sciences, mathematics, history, psychology, sociology etc. were written up by their colonizers, or by their sycophants - the native scholars. It continues even after decolonization. However, the Chinese government has been taking drastic measures to raise the Chinese nationalism. At the same time, high quality intellectual works are like good wine, which cannot be short-circuited in terms of time. Being a native of India, I must add that Indians are far behind the Chinese in raising their respective national resurgence.

The book has twelve chapters which are distributed over three sections - namely, Chinese Astronomical Canons and Calendars, Calculations, and Examples of Calculations. Wading through the first chapter transported me back in time - a year or two before my joining MA (Mathematics) in 1959 at Panjab University (PU), Chandigarh, where Astronomy, as a course of study, was replaced with Modern Algebra in MA Part II. It is worth noting that in the PU syllabus, there was a provision for the study of several subjects including Theory of Relativity, but there were not enough faculty to teach them. Curriculum diversity boils down to the economic health of a nation. In contrast, the US colleges offer buffets of courses and seminars sometimes even with 2-3 students.

Mathematics subtly trains the mind in a binary mode - either to go for a full grasp of the material at hand, or to just bypass it. That is how I felt about the material of the first two Chapters of Section I. Apart from too many texts and terminology in the Chinese language, it took me a while to understand as to why the time period of 104 BC - 1644 AD was selected.

The reference to Islamic astronics in Chapter 2 comes before the Hindu astronomy, which predates Islamic. This practice is very common in western academe - exaggerating Islamic scholarship and over the Hindu achievements. Also, it notes the influence of European astronomers with the arrival of Jesuit missionaries in China in the 16th century. Buddhism, a reformist movement of Hinduism, was introduced into China by the 3rd century BC in the Han dynasty lasting the 3rd century AD. Both China and India have used astronomical and astrological predictions in military expeditions and religious events.

Chapter 3 references to the Hindu ***navagraha*** - nine planets (Page 109). An interesting scenario tells how the astronomical tables were revised with the change of the national capital - from Nanjing to Beijing in the 15th century. The collective persona of political leaders pervades everything. Chapter 3 has a narrative on 'various zeros' both symbolically and conceptually. ***Kong*** in Chinese describes Zero and is denoted by a circle, O - a variant of the Indian dot. Also, it has a number of definitions and clarifications of calendrical terms explained in the context of modern times.

Chapter 4, one of the two small chapters having just 11 pages, is on the mean solar and lunar elements of all Chinese astronomical canons issued between 104 BC and 1644 AD. Chapter 5 deals with the calculations of moon phases with True Elements - concerning astronomical canons adopted between 619 and 1280. Chapter 6 deals with canons 1281-1364 and 1385- 1644. Mathematics involved in the two chapters suggest that the Chinese were quite sophisticated in the broad area of number theory. It eventually influenced astronomy in Korea and Japan. In 1732, the Chinese astronomical work was translated into French for the first time. Chapter 7 is devoted to the fundamental days of Mo and Mie in Chinese astronomy. At the end, it is shown how these concepts were mentioned in a classic Indian treatise, Arthashastra written by Chanakya in 300 BC!

The five chapters (8-12) of Section III, dealing with specific examples of calculations, imply the lack of mathematical theory behind them that a typical mathematician looks for. This, perhaps, may stem from the fact that the proofs, based on the Greek model of basic notions, definitions and postulates, had not captured the minds in China and India. By accepting Greek heritage, the Roman conquerors and subsequently the entire West simply ran with Greek heritage. Nevertheless, it should be understood that these examples are landmark astronomical problems and their results are interpreted and compared in Gregorian and Julian calendars, and other predictions like that of eclipses. These chapters are packed with several tables and long calculations on specific years, namely, 450, 451, 877 and 1417.

Finally, there is very little that I could distillate from the book, as far as hard-core modern mathematics was concerned. However, mathematical congruences are all over, which culminated in the famous Chinese Remainder Theorem in 1247, when the entire Europe was living in the dark ages. Also, there is not much that I could see sequentially in the name of the history of mathematics with Chinese emphasis. The entire material is heavily calculation based rather than having narratives on specific topics and themes.

May - June, 2017

COMMENTS

I haven't read the book, but from the sound of it, your review was much more thought out than the book itself. As you said at the end, a definite service to the profession! **George Buch**

You might like to see HR Gupta's work on perpetual calendar. Best regards, **Ajit Iqbal**

Hi Satish. I wanted to say that all this dumbing down of Hindu achievement in the sciences and the exaggeration of Islamic achievements in the same is all part and parcel of the politically correct appeasement of Islam. **Francis**

Dear Professor Bhatnagar, It was a great pleasure to read your review of the Chinese Astronomical treatise. Astronomy (both theoretical and observational) has been my favorite since teen-age. You have written an interesting and balanced review. **Arun Vaidya**

Hello Satish, Interesting as usual! My book on Indian Mathematics was reviewed recently in MAA. **George**

58. MATHEMATICS - ARCHEOLOGICALLY!

For the last few days, I have been trying to put my thoughts together for giving an answer to a real question - as to what mathematical notions did I discover in the ten-day (June 8-17, 2017) Chautauqua course, **Archaeology, Science and Mathematics in Ancient Greece?** Yes, this course was held in Greece- at various places renowned for their archeological sites. Besides the course director, twenty persons from the US and two ladies from Japan formed this study group. Almost all the participants had varied college teaching experience. About 4-5 of them were from mathematics. Nevertheless, the interest in Greek archeology and its ancient history had brought us all together.

The question of finding evidence and traces of science and mathematics in any part of the ancient world is not like that of a miner who goes out on an expedition for various minerals and precious metals on the riverbeds, deserts and mountains. **It should be understood that the tangibility of mathematics is beyond the sensory experiences.** Also, it is too simplistic to assume that one would hit upon something like a ***Lost Notebook of Ramanujan*** having hundreds of formulas and theorems, but without any proofs!

By the way, Ramanujan (1887-1920), not having the present luxury of paper sheets at his disposal in India then, used to work out mathematical details with chalk and erasable slate. This also provides a perspective when talking of mathematics in ancient Greece – as it means digging up history going back to the 9th century BC and even earlier. Naturally, my mind also ran on a parallel track in ancient India.

There is a plethora of material on any aspect of ancient Greece. However, the textbook material is pitifully isomorphic as far as the History of Mathematics (HoM) is concerned - whether of ancient Egypt, ancient Greece, ancient China, or ancient India. A side note - unless one has to refer to the books in languages other than English, one does not have to go overseas for research

in HoM. Inter-library loans are getting incredibly efficient – a book or a research article, not available, say, in a US library A, can be obtained from a US library B within 48 hours. This digital service is getting faster each year, as it is one of the metrics of the US libraries.

Also, one may go online even for visiting the museums and seeing their rare collections in showcases at one's own leisure and pace. The site of **Google Art and Culture** lets one literally walk through numerous museums digitally. A cursory look at this site tells that most major museums of the world are already online. The archeological sites are no different in this respect. At this rate, the day is not far off when one would be able to walk through the archeological labyrinths without leaving the comfort of home.

Despite these caveats, scholars would continue to go to Greece to learn about anything Greek. The reason is **the immersion approach of learning anything**. In our course, it included guided tours of the archeological sites and their associated museums, lectures, and simply soaking the Greek atmosphere in free time. One has to be able to extract and distill science and mathematics from the knowledge of the present too. It is what I call a **reverse excavation with an asterisk** though. Here, one faces a challenge and a dilemma of figuring out the ancient tools. Either, the ancient minds had them too, what we have them today, or whatever they had are lost forever. That is the bottom line.

For instance, the 4000 - year old Egyptian pyramids have been challenging us to figure them out for their very construction. Their builders had the scientific knowledge of seismic and geologic foundations, stone mining and cutting, horizontal and lateral transportation of 4000 Lb. massive stones, and the entire architectural blueprints. We continue to speculate on ancient technology. One of my delightful moments in the pursuit of mathematics is when a problem is solved in 2-3 entirely different ways. Therefore, for me, it is easier to apply this approach in other walks of life too. In particular, I accept the loss of ancient technologies that built the structures to have lasted for more than 2000 years.

Consequently, I do not accept the Darwinian or linear thinking on the uniform primitivism of the human race, as one goes back in eons of time. In my HoM courses, my thrust is on a line of thinking like this: if men like Buddha, Pythagoras and Confucius lived on this earth 2500 years ago, then such great men and women must have lived 5000 years ago too, and ad infinitum. Destruction of evidence by the forces of Time and Nature does not obviate their existence.

Acropolis and Parthenon in Athens epitomize not only art and architecture, but science and mathematics too. For instance, the mathematical Golden Ratio is there in certain planes. However, in order to make the Parthenon look 360-degree aesthetically beautiful, the Golden ratio is downplayed in some planes. Parthenon is perfect due to or despite its mathematical 'imperfections'. Every year, landslides in the USA, for various reasons, simply obliterate homes, but the Pantheon has been sitting royally on a huge rocky bed for millennia. It is a testimony to the sophisticated knowledge of math and engineering that the ancient Greeks had.

After spending three mornings outside in Acropolis, we spent two hours inside the National Archeology Museum. James Powell's talk on **The Antikythera Device** - used by the ancient Greeks provided evidence on the existence of technology in ancient times. Moreover, a recent publication (both in the Greek and English languages), ***Inventions in Ancient Greece*** by Kostas Kotsanas, has a list of nearly 100 of them. I just ordered this book from Amazon for the sole purpose of researching a possible connection with inventions in ancient India. The golden periods of both Greece and India fall identically in the BC era! Furthermore, Alexander, the Great's outpost in western India survived for several decades.

The archeological site in Delphi, 180 KM from Athens, was another stunner. Nearly a dozen pillars of the temple of Apollo are still standing after 2000 years, so are the amphitheater and stadium - all on one slope of a mountain, which also provided all the rocks needed for various structures – they are still breathtaking. It is a clear testimony to the embedded geoscience, engineering and math in ancient Greece. The museum attached to this archeological site is right on the foot of the mountain. The immersion approach worked on my mind effectively. The guide for the first two sites was quite knowledgeable.

The third archeological site was once a full-fledged city on the island of Delos. We ferried there from the island of Mykonos. No one lives overnight on Delos - its current size is hardly two square miles. Existing sections of its underground sewer system, marble pillars, mosaic colored tiles and marble statues, and an amphitheater, though worn out after 2500-3000 years, nevertheless, speak loudly of its bygone splendor as a strategic trade location, and much more. No one has to convince me about the knowledge of science and mathematics the ancients had. We ran out of time for a small associated museum of Delos, which houses many precious items removed from the open areas.

The glorious days of Delos reminded me of the present emirate of Dubai, which has been developing since the 1970s, as the number one international destination for doing any kind of business. However, I must add that the technological wonders of Dubai have no connection with the contributions in math and science of the native Arabs of Dubai. Dubai did not even have a college until 20 years ago! Delos has a similar story.

Archeology gives a unique perception of the human timeline. One can see how in a couple of millennia, the rivers change courses, mountains erode; earthquakes, tsunamis and volcanic eruptions desolate a humming community - apart from the regional socio-political revolutions. The human lifespan is literally reduced to a dot! In this context, geology takes one to a limit on earth; however, modern astronomy really takes me to the ultimate physical limits in the cosmos. I enjoy such flights of imagination. They are triggered during wanderings in a maze of an archeological site.

The ancient town of Theran (Akrotiri), on the island of Santorini, remained buried under the volcanic ashes and debris for over 3600 years – the result of an epochal volcanic eruption in 1627 BC! Its accidental discovery and subsequent excavation, by French archeologists, has been going on since 1967 to form a union of modern science and technology with that of antiquity! The excavation of the site is done by choosing earth 2 cm in depth at a time. This attention to the detail in unearthing in archeology is like a code in a computer program or steps in a mathematical proof! I relished this thought of mine.

Interestingly enough, this archeological wonder has created a new tourist town with a bustling economy. Precious items found in excavation are carted away and preserved in the museums. Looking at those restored pieces compelled me to think that the pursuits of science and mathematics are also a corollary of a materialistic and sensory way of life – having no limits on sensory delights. One can see this in the USA today. It must be noted that art and literature follow science and mathematics – like pioneers and settlers in a new land!

The tour of the **Palace at Knossos** in Heraklion/Herculean on the island of Create lived up to its name. It validates my hypothesis that science and math are embedded in the monuments. Just think of this palace complex dating back to 1380—1100 BC, built on a bed of rocks, 280' above the sea level. The same archeological story – that it was accidentally discovered in 1878, and its excavation was charged to an English archeologist for 35 years. Simply put - whatever it would take to build a palace today, the ancients did it 3000 years ago. Period.

Its associated museum is on a street across from our hotel, **Capsis Astoria**. It houses all the precious findings from the sites. People, like us, come from all over the world, stay in the hotels, eat in the restaurants, and buy souvenirs – all this generates an economy of its own kind. Thus, a city is also re-incarnated!

From the point of view of the HoM, I have two conclusions. In a region, large enough, there is a section about which one would say indisputably that science and math flourished there. And, not too far from it, there would be another section, where its people had no faith in math and science, but occult and superstitions prevailed there. Specifically, I tell my HoM students to imagine Las Vegas valley after all kinds of natural and man-made upheavals taking place in the next 3000 years. If archeologists excavate a site, say, where Caesars Palace is sitting today, then they would conclude that Las Vegas valley was very advanced in science and math. But, they would conclude the valley as primitive in math and science, if they excavated a northwest section of downtown Las Vegas. The bottom line is that any archeological work is very expensive, but the tendency to generalize from meager findings is irresistible. I constantly caution my HoM students.

Without such archeological trips, I would not have drawn this conclusion – this time five sites were visited in ten days. At the end, I really felt saturated – like one feels after a Las Vegas dinner buffet. Also, for the last one month,

I have been reading off and on, ***Ancient Greece*** (2013) by Thomas Martin. For me, Greek immersion has been a totally rewarding experience.

June 28, 2017

PS: I am working with the Director of Ancient Explorations for having a similar Chautauqua course on **Archeology, Science and Mathematics in Ancient India.**

COMMENTS

Thank you so much for sharing your thoughts and overview of the trip with us! It is a great reminder of all the wonders that we were able to experience. It was nice traveling with you. Best regards, **Cynthia Huffman, Math Professor, CSU LA**

Glad you were able to truly enjoy the ruins. Math and time are key to ancient Maya culture as well. **Diane Chase**, UNLV Provost

Love it...although I am biased because I live everything Greek. Have you ever been to Hagia Sophia in Istanbul? **George Buch**, UNLV Math Lecturer

Satish, Thank you for the paper. What a great journey! I am jealous. I would have loved to have visited the sites to confirm that human civilizations would be meaningless without mathematics. It is the foundation of not only science, but also social science (commerce, economics, and yes, politics! Mathematics is integral to the arts and music. I see your point of view about the article. We, as seniors, have so little time to waste! I get caught up in the minutia. I would like to sit and hear more about the Chautauqua course. **Michael Luesebrink**, UNLV's science and math librarian

Dear Satish, Awesome reading! Have finally had time to relax and enjoy savoring this beautiful article. Loved your reaction to and interpretation of the "mysteries of Ancient Greek civilization revealed"! You are an anthropologist at heart, you know! And best of luck on your Chautauqua! Wow! **Mannetta**, free-lance anthropologist and UNLV benefactor

Satish, we appreciate the invitation. Read your extended reflections on the trip. Need to read them again. Your thinking is so very deep and profound. You are a natural contemplative! Attaching several photos of places, we shared. Hope to travel with you again; **Best, Jim and Anne**

59. OLD-FASHIONED STUDENT CHEATING DYING!?

Today, I nervously walked to the classroom to give the Final Exam in a course that was finished last week. The reason was as to how would I stop the students from cheating and looking over each other's solutions? Well, let me set the stage for my concern. The sloping classroom has exactly 60 seats divided into two isles of 30 each. The number of students being 51, they normally sit 3-4 inches apart from each other.

One may ask as to how I handled this situation at the times of the class tests during the semester? There were two helpful factors. Number One - I was provided two student proctors, who watched the students for any cheating tendencies or efforts. Number Two - the two tests had identical problems, but in different orders. Thus, no adjacent students had the same tests. For grading convenience, one test had white sheets, and the other yellowish. The performance of the students in class tests did not suggest any cheating.

However, today, I was all alone - no proctor was available. Besides, all 51 students had the same tests and the same order of problems. There being a rush on the office copier to run the tests, I simply forgot to permute the problems! Well, I distributed the tests and immediately, my eyes became hawkish for catching any test infractions. Amazingly, for a full first hour, no student even furtively glanced at their neighbors' solutions! Let me add that the answers were not based on multiple-choice questions, but were rather descriptive in nature. Now, it became an inverse problem for me 'why was there no cheating?'

In India, the country of my national origin, cheating in exams, particularly, during the Finals becomes a law and order problem. Having taught in Malaysia (in Southeast Asia) and Oman (in the Middle East) the students are brazen in copying and plagiarizing. Are the US students different? Yes, to some extent. My one guess is that about 50% of the students in this course are first generation immigrants. The US, as a proverbial melting pot, is no longer the same. During my graduate student days (50 years ago) at Indiana University, no student helped others or sought help on homework or take-home exam problems. Instructors seldom watched over the students taking the tests.

Getting back to finding an explanation for no cheating by these students: when you focus all your energies on a problem, then its solution is bound to crack open. In mere ten years, smartphone technology has discretized the kids (from the age of three!), the teens, and the young adults. On the campus, students are seen walking while watching, reading, laughing or listening on their smartphone - absolutely impervious to the people around. This state of mind of being all-able-to-know-anything without any human help has extreme sociological and psychological implications. At least one positive side of the new 'island identity' is that the tendency to cheat is minimized!

In any cheating scenario, a 'cheatee' may be helping a cheater on some idealist principle. Equally, a cheater may rationalize his/her actions on some social grounds. However, this helping nature is not instinctive. It only grows with time in social interactions. The i-technology is galloping in the directions of robots, driverless cars, and pilotless planes (drones) - thus, driving humans out of human lives!

My second hour of watching the students was mainly spent in writing out the solutions of the exam problems. Nevertheless, all through the hour, I felt good at my 'discovery'!

December 11, 2017

COMMENTS

That's a good group you have. I let my students use a note sheet for in class exams...has ended all occurrences of cheating because they have what they think they need right in front of them. **George B**

Very interesting observation indeed! Here in India too, every student has a smartphone and so perhaps the same thing happens at examinations, but no one seems to have noticed! Of course having retired 22 years ago, I Have no first hand experience but I'll find out from others. **ARUN VAIDYA,**

Dr. Bhatnagar, I do have the same observation! They simply don't cheat. I had one this semester using a phone calculator on his lap, but I told him he was permitted to use it anyways, as it was a pop-quiz!
Sincerely, **Andrzej Lenard**

From personal experience, my cheating days ended after high school. I only cheated once in a while and only when I really needed to, but this is not the point. The reason why I cheated at all was because I didn't have the skills or resources to get the solutions myself. Now, there are plenty of websites that can guide us through specific problems, but these services are rare and not popular yet. The reason why I stopped cheating going into college was because I became more independent. I learned to read highly technical material. I became efficient at researching. There was no reason for me to cheat anymore.

There's no such thing as "I can't figure this out." Plus, you have to take into consideration that college students have more pride in their work than high school students. Most of us chose to go to college, not forced. So in a sense, we're here because we want to. I want to be good at what I do. If I cheat my way through college then the degree is nothing more than a piece of paper.

Now regarding technology's impact, like you said, there are advantages and disadvantages. As I mentioned, one of the advantages is that we're more independent, but one of the disadvantages is that we're too independent. Sometimes I spend too much time trying to learn by myself instead of reaching out for help. Sometimes there are projects that should be done by a group instead of an individual, to which I fail to acknowledge. Best regards, **Sang Huynh**

Cheating is an interesting phenomena – a philosophical issue at heart. I think you are more than on to something and I would even suggest you have a sociological research project. I would encourage you to get together with a social scientist, probably a sociologist to set up an experimental protocol to refine your hypothesis and test it. It would be publishable in any number of journals including Ed psych. Cheating has been an eternal universal problem. When I taught an accounting course in Moldova after the soviet union broke down, I was amazed at the blatant cheating that occurred until it hit me that communism encouraged group-thinking. A thought---**Bob moore**

60. LIMINALITY THROUGH A MATHEMATICAL PRISM

[**Note:** This reflection has remote connections with the history of mathematics, but I felt its inclusion will add newness. On February 25, 2018, I presented portions of this article in an interdisciplinary panel on the esoteric topic, **Death, Liminality, and Cultural Expression**. The occasion was the 30th annual conference of the **Far West Popular Culture Association** held over the weekend of February 23 - 25.]

Etymology of Liminality

The range of human experiences is mathematically infinite. No language has a vocabulary vast enough to capture them all - even for a virile language like English, wherein, the vocabulary is always changing, as the new words are constantly added every year. A quick online check on Dictionary.com spelled out the following meanings: 'the **transitional period** or **phase** of a rite of passage' and 'the condition of being **on a threshold** or at the **beginning of a process**' (which implies the **end** of another process) Liminality is not an heightened moment, but it does have an experiential dimension to it.

Mathematical and Historical Touches

In contrast to Liminality, 'Death' in the title is very familiar - both as a word and a concept. After all, what is Death? So much has been written and experienced on Death in every religion, culture, society, and even in the academe. At the age of 80-, my perception of Death has jelled in as an irreversible state of a being. With this definition, everything is dying in every moment - the threshold of Liminality can be perceived as short or as thin as possible - like the positive real numbers in mathematics, which do not have the least positive number. That was an aha moment for my participation.

Liminality simply captures the state of transition, where the contextual time duration is relativistic. Thus, the word Liminality may be new, but the associated human conditions are absolutely not new. The coinage of the word, Liminality, reminds me of the common phrase - 'old wine in new bottles'. At this point, I must interject that after 55+ years of teaching, researching, and writing about mathematics, mathematics has become a major paradigm of my life. It is my favorite window, a prism through which I see life. I enjoy, understand, and try to explain every aspect of life through mathematics. However, realizing the ultimate limitations of any medium of communication, lately, I have started working on a kind of 'quantum' thinking too, as opposed to the deductive thinking of mathematics.

Liminality in Mathematics

In a sense, this article has been written as an intellectual roller coaster of its kind. From this point onwards, its subsequent format is borrowed from a stray handout on **Thirty-six kinds of Proofs**, which was received 8-10 years ago.

Anyway, the first mathematical liminality is encountered as a rite of passage in a middle school, when a student moves from the comfort zones of arithmetic of natural numbers and geometry of circles, triangles, and rectangles to variables of algebra. Suddenly, the notation like xy does not have the same meaning as a number, 34. Also, while expressions like, 3 + 4 = 7 make sense, x + y stares at you for a corresponding meaning. It takes a while to understand the dances of the numbers and variables.

The second significant mathematical liminality is faced in high schools or during the first year of college, when one moves from algebra, trigonometry, analytic geometry into the realm of calculus, where one struggles with limits and infinity, which is not the largest number! For math and computer science majors, a proof course, like Math 251 (Discrete Mathematics I) at UNLV, captures the essence of Liminality in mathematical development of the students. The courses taken before a proof course begin to look different from the ones taken afterwards.

Liminality in Sciences

Biological growth, observed externally can be described as a series of liminical phases - from childhood to adulthood - internally, from a unicellular system to a very complex system of systems. An example is of a human body as it appears in adulthood.

Liminality is a question of the focus of the mind at a particular instant. The mind pretends to ignore the phases before and after the transient state.

But these three states are 'mathematically continuous' - thus, inseparable! Therefore, the following theorems in mathematics and principles in the sciences capture Liminality, as they are the exclamation points of the transient periods preceding them.

Godel Incompleteness Theorem, Fermat's Last Theorem, Critical point of a real-valued function, and mathematical limit, of a function, say, like, $\frac{Sin\,x}{x}$ as x approaches 0. (limiting process captures Liminality)

Kepler's Laws of planetary motion; Newton's Laws of motion and gravitation were formulated by Newton, when he returned to Cambridge University after spending two years (1665-67) in the countryside - away from the Great Plague of London, Einstein's Special Theory of Relativity (1905), the world's most recognizable formulas, $e = mc^2$ and General Theory of Relativity (1915). Liminality is hidden in the incubation period of these formulas and principles. In the field of psychology, Freudian Psychoanalysis is a liminal milestone in psychology.

Nature and Liminality

The flow of water when observed from a fixed point and a saying, 'you never step into the same river water again'. The eerie silence before a storm, avalanche or earth movement, the falling of the leaves in the fall season and making the trees look like perfect skeletons, and the dormant period before the sprouting of new leaves captures liminality.

Hindu Mythology

The Hindu Trinity of God, namely, Brahma, Vishnu and Mahesh symbolize birth, sustenance and dissolution in the cosmos. Sustenance is Liminality. It is captured in a popular story of Prahlad and his father, Emperor Harnakshyap, whose being killed is a perfect example of Liminality. Amusingly, Trinity is built in the word, GOD, where G is for Generator, O for Operator and D for Destroyer!

Astronomy

Liminality is constantly seen in the universe - from its vastness where galaxies come into existence, thrive for 'a while'; and then extinguish in some black holes.

Religious Rituals

Baptization, the accepting of Christ as Savior, Born Again, radicalisation, the making of the terrorists - all are the plays of Liminality - thresholds in human life spans. The focus is not on a particular ceremony, but on a period of inner transformation that leads to the external event.

Violence in Urban Culture

Stephen Paddock, the Mandalay Bay shooter - his life before the shoot-out, during ten minutes of the firing(10:05 - 10:15 PM) on 10/01/2017, and the ending of his life with a self-inflicted gunshot in his mouth. Similarly, Nikol Cruz's carnage lasted six minutes - killing 17 students and wounding a dozen. They are also instances of Liminality in life.

Micro Human Sperms

On the average, each male ejaculation releases nearly 100-400 million sperms. The life cycle of the sperms in the male body is approximately 75 days. Once they are out of the body, they are on a threshold of life and death, which may last from 20 minutes to 200 minutes - depending on various factors, -like, night discharge, masturbation, and sexual penetration.

Induced Liminality

The mind-altering drugs, like opiods, can create states of Liminality.

Liminality in History

The global wars like World Wars I and II, and revolutions like those of French, American, Russian, and Chinese capture Liminality during the periods that they last. The Nazi propaganda machine during World War II provides a an historic scenario of liminality in modern communication. The life spans of great individuals - like, Alexander, Buddha, Genghis Khan, Confucius, Jesus and Gandhi are liminal scenarios, as the societies have been transformed by these greats.

Inventions like those of iphones, ipads, ibooks etc. are liminal landmarks in the sense that the time it took for these inventos sets the life before and after poles apart. It is resulting in discretization of societies.

Graduation Ceremonies in which a student walks up to a podium, receives a diploma, and then walks down, is an event which captures a liminality in an individual of its own kind.

Geographical Liminality

Every day, the twilight and dusk illustrate liminality. Equinoxes and solstices are perfect examples of annual Liminality. Between New Year Eve and New Year Day, where is Liminality? In the Time Zones, say in the US, Liminality here is an invisible line that divides one Time zone from another. The border between countries is Liminality. One can only experience this Liminality while crossing the US - Mexico border.

Death and Reincarnation

Bardo, is a Tibetan word, which means the phase, where the human 'soul' is suspended before taking another birth. The ***Tibetan Book of the Dead*** is an excellent source for its study.

Liminality in Socio-political Events

The acquittal of four cops in the Rodney King trial (April 29, 1993) unleashed ethnic riots. And, the acquittal of OJ Simpson (Oct 03, 1995) triggered ethnic euphoria after a trial that lasted for 16 months. A person undergoes Liminality in the detention center of immigration service or in the quarantine area for the sick.

Sports

The drafting periods of 1-3 days in major professional sports in the US are perfect examples of Liminality. Some 18-22 years old athletes living in poverty become instant millionaires, the moment their phones ring and they get the news of their being drafted by some teams. For instance, Lebron James lived in dire poverty, but in June 2003, his fate changed. Today, he is the first billionaire in all professional sports - amongst men and women.

Sexual Orgasm

In my paper presentation, sexual orgasm during intercourse was the last example of Liminality. It is captured during the male ejaculation. For a few moments, it is the beginning of the shutdown of all the five sensory gates for a few seconds.

In summation, any kind of quantum leap - up or down, underscores a liminal experience, which remains subjective. Above all, Liminality is happening in every physical, mental, and psychological phenomena in the universe all the time. Its perception depends on the sensitivity of one's mind.

March 04, 2018

COMMENTS

Interesting. I remember GOD: Generator, Operator & Destroyer from last week's ***Bhagwat katha*** by Sri Lal Govindji. Warm regards. **Girija**

Thank you, Satish! You have given me such a pleasure that this topic, liminality and death, has visited the muse upon you in a manner that has brought you joy and the creative urge. May her stay not be liminal. Till next year, Warm regards**, Roberta**

61. HISTORY OF MATH DEFINES NATIONS

"The School of Mathematics and Statistics at the University of St Andrews is known internationally as the home of the MacTutor History of Mathematics Archive; appointment of a Reader/Senior Lecturer in the History of Mathematics. Salary: £51,630 - £58,089 per annum, Start: 1 January 2020 or as soon as possible. Candidates must hold a PhD in History of Mathematics, or Mathematics, or a cognate discipline. details can be found at **http://www.st-andrews.ac.uk/"**

The above paragraph has the essential extracts of the advertisement that was forwarded to me yesterday by the SIGMAA-HOM (Special Interest Group of the Mathematical Association of America in History of Mathematics). I have been a member of the MAA for nearly fifty years. As soon as I finished reading the position advertisement, the following thoughts raced through my mind in the context of HoM in India, UK and the USA:

To the best of my knowledge, the HoM has not even found a foothold in any undergraduate or graduate curriculum of an Indian university. According to an online source, there are nearly 800 universities (non-comprehensive as compared with the US universities) and 40,000 colleges in India. My guess is that 90% of these institutions did not exist before 1947 when erstwhile India was freed while one-third of its territory went into the creation of the new Islamic nations of Pakistan and Bangladesh.

No specific institutional comparisons are intended here. The US has nearly 5300 colleges and universities. It is worth noting that the population of India is more than three times that of the US modulo factors of literacy, economic conditions and cultural history.

The existence of degree programs and courses (papers) on the history of a specific topic - be it in Math, Science, Arts, Internet, Music, Psychology etc. is a measure of the intellectual depth and breadth of that particular society. India has been a free nation for 72 years, but it has a long way to go for the holistic intellectual development of its public as well as its intelligentsia. In democratic India, the decolonization of mind-sets is a much longer process. Indegenous studies of history in general and that of

HoM in particular are subservient to the study of science and technology. In India, the current education is highly fragmented and furthermore, it forces early specialization at the cost of humanistic development of the young minds.

According to a note in the ad, this is the first such faculty position created in the University of St. Andrews, which was founded in 1413 - almost 500 years ago. By way of perspective, Oxford is the oldest university in the UK - founded in 1063. Five years ago, one of my MS students researched the existence of doctoral programs in HoM in the US. He identified the University of Virginia as being the only one. In contrast to the US, the UK has far more academic institutions where one can pursue advanced degrees in the HoM. One may raise the question as to why this is so. What follows is worthy of further investigation.

The UK being the greatest colonizer in/of the world, the objective of the education systems as implemented in the colonies was to glorify the British ethos through the teaching of the history of every aspect of British life. Most founding fathers of the US, being of British ancestry, were cognizant of the British designs of education in thirteen colonies. Soon after independence, they surgically cut the colonial education system out and started developing their own models of education. Today, Harvard, Yale and Cornell, to name a few, are the benchmark universities that the world looks up to.

In order to promote HoM, I have endowed two awards in HoM viewed as a subset of humanities - including Mathematics, History, Political Science, Religion, Archeology and Sanskrit. These awards are administered by the Indian Mathematical Society of which I have been a life member since 1980. It is a small step in the direction of renaissance of history in India.

June 21, 2019 (Bathinda)

62. IN SEARCH OF INFINITY

Yes, this is also the title of the book which was first written in the Russian language by N. Ya. Vilenkin in 1995, and recently translated into English. Two days ago, it was placed into my hands as an exchange gift from my Russian colleague who is leaving for a Connecticut university after five years at UNLV. Having it being given in such a thoughtful manner, I could not help browsing it right away. Otherwise, my desire to read has diminished, as my time has come to give back to society through my own writings.

The title is tantalizing. After all, 'Infinity' is a thing; yet, 'Infinity' is not a thing - as is a book, donkey, Sanskrit, number 7, and rose flower etc. - all solid and earthly entities! The seed of Infinity lies in the wonderment felt by a person who gazes at the stars, watches ocean waves, feels for sand particles amidst sand dunes, and looks at leaves of trees in a grove. Infinity belongs to all thinking persons - be they poets, astronomers, monks, microbiologists, economists, or philosophers. I tend to think that curiosity about Infinity and the desire to understand it distinguish humans from other species. Who knows that one day it may be discovered that certain other species have their senses of Infinity - a realm of fictional neurology until then.

As a matter of fact, Infinity is like the proverbial elephant that is studied by a bunch of blind (one-dimensional) persons. For the sake of simplicity, let all the approaches for grasping Infinity be classified in two categories - mathematical and non-mathematical. For instance, a performing musician reaches out to his/her 'Infinity' that even the audience can immerse into that crescendo. It is true for an artist, writer and other seekers of Infinity.

However, the focus of this book is to understand 'Infinity' solely through the framework of mathematics and physics. In a short note, the Editor-in-Chief states that the book is about 'the relevant contribution of Russian scientists'. Right away, this raised a question in my mind - what about the contributions of Indians in their understanding of 'Infinity' ? Here below, is my brief, yet wholesome narrative on this subject:

The Sanskrit ***mantras*** are energized syllables which are very condensed in writing. Generally, multiple meanings are embedded into them. They are the outpourings of the holistic minds (called ***rishis***) in their yogic states. Given below is one of the dozen Sanskrit mantras invoking peace and harmony in the universe. It is worth noting that one of its meanings is a clear characterization of Infinity as stated in mathematics.

ॐ पूर्णमदः पूर्णमिदं पूर्णात्पूर्णमुदच्यते ।
पूर्णस्य पूर्णमादाय पूर्णमेवावशिष्यते ॥English transliteration:
Om Puurnnam-Adah Puurnnam-Idam Puurnnaat-Puurnnam-Udacyate |
Puurnnasya Puurnnam-Aadaaya Puurnnam-Eva-Avashissyate ||
One meaning:
1: Om, That is Purna (Full) This (Inner World) is also Purna: From Purna is manifested Purna,
2: Taking Purna from Purna, Purna indeed remains.

This ***mantra*** is drawn from the Mandukya upanishad, one of the several upanishads dating back at least 2000 years. This example is one of the tips of icebergs - knowledge lost, systematically destroyed, and now forgotten. When a society, at large, gets obsessed with abstract thoughts, then it loses its fighting spirit. Eventually, the enemies who have infiltrated within and the so-called barbarians from without, take over the peaceful and prosperous society. For India, the 13th century was the beginning of never-ending invasions and immigration of hundreds of foreign invaders who decimated the ancient heritage of the Hindu India, which continued through the middle of the 20th century. As a historian and political meteorologist, I can now see the down-turning of the USA that started almost fifty years ago, but it has been gaining momentum lately.

This reflection is not a book review, as I have not read even ten pages out of a total of 140 pages. As far as the foundations of mathematics and physics are concerned, one simply has to go to the square one and really contemplate over the definition of a point (no length, no breadth, no width - that is it; nothing about what a point has!) in mathematics as Euclid gave it 2300 years ago, and the definition of a particle (defined as a Euclidean point where, its mass could be anything from 1 gm to1 ton. How is that?) in modern physics (since Newton). Despite such 'holes' in the foundations of mathematics and science, men have landed on the moon with precision. This is no less paradoxical to me!

If there is one concept that sets the world of mathematics apart from the rest of the disciplines is the notion of Continuity at a point that is built on the definition of Limit at a point, the acme of collective thought of mankind since the time of great Greek minds. The physical world is all discrete in every divisible state. However, nothing comes closer to the continuum in mathematics. Also, my considered opinion is that the one-to-one and onto functions, the fancy names of counting, have been two-edged swords. They do clear the fog on one side but create new paradoxes at the same time.

In summary, the march of the collective and collaborative body of intellectual knowledge is essentially based on binary logic (call it rational thinking), which, for most humans, is drawn mainly from the left side of the brain. The fundamental question is about the existence of a non-binary logic or non-rational thinking. Let me emphasize that non-rational thinking is not irrational thinking which is filled with contradictions. It is the same end that matters both in rational as well as in non-rational thinking.

Here is an example of non-rational thinking. My autistic grandnephew, when around the age of 10-12 years, used to complete a jigsaw puzzle of 500 pieces within an hour. He cannot explain how he does it and we have no way of figuring it out either. Is he using the right side of the brain to solve problems? We, the so-called normal people, can't harness the mental skills of autistic and such other people. A key for a new understanding of Infinity may lie with the thinking of such marginalized persons.

Meditation is also described as the awakening of both sides of the brain. It culminates in enlightenment - the understanding of 'Infinity' holistically. In conclusion, Infinity is such a beast that 80- 90 % of the world's population is not bothered about it, some are scared, some are so confused that they equate this symbol, ∞, for Infinity with Infinity! Nevertheless, a minuscule part of the population - including myself, get high on Infinity. The set theoreticians amongst mathematicians thrive on Infinity professionally, as they have an entire universe of infinities! These infinities in mathematics are like infinitely many gods and goddesses in the Hindu pantheon - due to the underlying assumption that divinity (life force or atomic energy) is embedded in every entity!

The bottom line (mission of life) is to find your own Infinity. Actualization of one's potential is one path. That is called Self-Realization in the yogic lexicon - and that it becomes a window on the universe - according to the sages of ancient India! Nevertheless, the search for 'Infinity' is a pathless path.

July 17, 2019

COMMENTS

Very well articulated with remarkable brevity **SR Wadhwa**

In the late 1940's, Professor Raymond A. Lyttleton proposed that the structure of the human brain and the structure of the Universe were somehow similar. This seems to tally with your relationship to Infinity regarding the use of the two hemispheres of the brain. Perhaps the Universe is infinite. Then there is the strange phenomenon of physics whereby a particle does not settle in one place until it is observed and measured. There is clearly a connection between conscience and quantum physics.

With regard to autistic children, it would seem that what we consider as being mentally 'normal' and 'abnormal' may all be a matter of cultural convention. Could it not be that what the culture refers to as 'mental illness' may well just be an alternative way of seeing reality - whatever that is? **Francis Andrew**

Probably based on the Vedic thought Gurubani repeatedly talks of 'Infinity' as Akal Purakh. The shloka 'Om Puurnnam-Adah Puurnnam-Idam Puurnnaat-Puurnnam-Udacyate.....' preludes the thought repeated in 'The Adi Granth'. Your Reflection and the 'Infinity' through the framework and tools of mathematics and physics' is remarkable. Bravo!! Please forgive me if I 'over emphasized' anything. **Jagjeet Sahney**

Professor Bhatnagar: I admire you for tackling such esoteric subjects as Infinity. How far does space extend? When and how did God decide to create the cosmos? Where are all the souls that have been created since then? What happens if a fraction of our galaxy (hopefully some part far from us) is swallowed by a black hole? Keep it up. **Gurnam S.S. Brard**

Love this Reflection. Thanks for sharing. **Rahul**

Dear Dr. Bhatnagar, Sorry for the delayed response! Was pleased to receive and read the reflection. Enjoy the reading, agree on the "pathless path". Wish you to have a smooth Summer 3 session and an enjoyable rest of this summer. Hope to stay in touch. Sincerely, **Viktoria**

63. HUMANISTIC HISTORY OF MATHEMATICS

What is the Humanistic **History of Mathematics**? I had never heard of it before my thoughts started converging on this topic two months ago - even though I have been attending the annual Joint Mathematics Meetings (JMM) for the last couple of decades. Moreover, the sessions on **Humanistic Mathematics** at the JMM have been around for at least three decades. The papers presented in the sessions have a very broad appeal. Over the years, I have enjoyed presenting papers in the sessions of **Humanistic Mathematics.**

Partly, because of the popularity of these sessions, the first ***Journal of Humanistic Mathematics*** was eventually launched in 2011 - may be the first of its kind in the world. Continuing on the thread of 'Humanism' in mathematics, years ago, UNLV had an elective course, ***Humane Mathematics*** (Math 101). Yes, mathematicians and mathematics are two different entities! Mathematics is value-free - indeed the most objective of all disciplines. Anyway, this course was like one of the 101 introductory courses offered in most US colleges - this one was on the appreciation of mathematics. I taught it a few times before it was deleted unceremoniously. Sadly, there is no such math course in the catalog.

As far as my teaching of History of Mathematics (HoM) is concerned, apart from an experimental undergraduate course taught in the mid 1980s, my interest in HoM has been evolutionary in nature. It was gradually embedded in my growing interest in history in general. The turning point came in 2002, when the fourth master's concentration, **Teaching Mathematics,** was approved. One of three required courses in this graduate program (master's only) is **History of Mathematics** (MAT 714).

It was in Spring-2007 that I taught MAT 714 for the first time, and since then it has been done every two years. The textbooks on HoM being almost isomorphic, in a one-semester course, any approach towards the teaching of HoM does not deviate too far from textbooks. However, the course description provides a lot of flexibility depending on the instructor's expertise, and mine is humanistic.

A sidebar: in April, 2015, I organized the first session on HoM in the Western Section of the American Mathematical Society. The theme of the session was essentially the **Necessary and Sufficient Conditions for the Development of Mathematics.** It was quite an innovative and challenging topic, and it drew a few diverse and interesting papers.

Once I decided to approach HoM as a subset of History in general, then there was no stopping at History! In subsequent offerings, I included flavors of Economics, Philosophy, Politics/Political Science, and even organized Religions too, as they clearly impact HoM. Science serves like background music to HoM. During a narrative, it is enlightening to notice that a specific political regime boosted science and mathematics in a particular era. That is how **Humanistic HoM** defines itself. And, I consider myself a **Humanistic Historian of Mathematics**!

September 09, 2019

64. BLACK HISTORY MONTH & MATHEMATICS

History of Mathematics (Math314) is a unique course in many ways. What I love about it is that every once in a while, I present class material at the spur of the moment - that is not connected with the textbook or with anything done in a previous lecture. In most undergraduate math courses, neither it is advisable nor possible to deviate from the syllabus that is often very tightly strung. Moreover, the contents being linked to other math courses, the coverage in a typical math course has to be done rigorously.

Last Monday (02/24), while walking on the westside academic mall of the UNLV campus, I stopped by two young Afro-American (AA) students manning a table of their club. On my asking, if they knew any significance of the month of February, they looked nonplussed. When I told them about it being the Black History month, one simply nodded about it. **The point is that if you don't care to know even a bit of the history of your race, then who else would care about it.**

Let us first be clear who the AA are in the US. **In mathematics, the definitions are very precise and unambiguous.** However, in the US, Afro-Americans are those citizens whose ancestors were shipped from Africa as slaves in the US. Thus, every black person is not an AA. Likewise, Africans who immigrated to the US after 1865, when slavery was abolished, do not 'qualify' to be AAs. In practice, this definition becomes contentious - that is the nature of the world outside mathematics!

The Negro History Week was a precursor to the Black History Month. The Week was created in 1926, when historian Carter G. Woodson (1875-1950) and the Association for the Study of Negro Life and History announced it to be the second week of February. This week was chosen because it contained the birthdays of Abraham Lincoln (February 12) and of Frederick Douglass (February 14), both of the dates black communities had celebrated together since the late 19th century. However, it remained limited to a couple of southern states.

In 1976, Black History Month started being celebrated all across the country, when President Gerald Ford recognized it, during the celebration of the United States Bicentennial.

With this background, I decided to work the history of mathematics into the 2020 - February of which only five days were left. I quickly prepared a six-question quiz-cum-survey to gauge students' awareness about the Afro-American history, and contributions of the AA in mathematics. Twenty-nine students out of a class of 33 filled out the questionnaire and I just tabulated their raw responses, as given below with my questions:

1. The name or description of an AA teacher that you had in **elementary** school? 11/29 (means 10 out of 29)
2. The name or description of an AA teacher that you had in **middle** school? 14/29.
3. The name or description of an AA teacher that you had in **high** school? 18/29
4. The name or description of an AA teacher that you had in **college**? 19/29
5. The name or description of any AA mathematician that you have known or heard? (5+5)/29
6. Anything you know about the contributions of AA in the history of math and science? (10+5)/29.

I do not see any point in doing any statistical analysis on this sample - except, to explain the first numbers - namely, 5 and 10, in the parentheses pertaining to Question # 5 and 6. These responses are based upon a popular movie, ***Hidden Figures*** (2016), on the lives of three AA women who worked as mathematicians and/or engineers in NASA's Apollo mission in the 1960s and 70s. What a coincidence that Katherine Johnsom died on 02/24/2020 when I talked about her in my class! She was 101 years old. The point behind these parenthetical numbers is that it clearly tells how much 'learning' takes place through the entertainment industry and social media - far beyond the confines of the academe.

After the 'quiz' was collected, I spent the rest of the period discussing the life and times of Benjamin Banneker (1731-1806) and a fantastic revised book (1999) on his life, as written by Silvio Bedini, an investigative writer. Banneker was a free slave, but he was the grandson of a white woman from England who was sentenced to exile in the US as an indentured slave (five-year bond). Banneker was a polymath - an applied mathematician in the modern classification, astronomer, poet, artist, maker of a clock besides furniture, DC surveyor and a social passivist. I dwelt on the states of mathematics and colleges in the colonies before the Declaration of Independence in 1776. To the best of my knowledge, Banneker was the first Afro-American polymath, a multi-dimensional intellectual - the type of people I admire.

On personal AA experiences, in 1968, I came from India to Indiana University, Bloomington for getting an American PhD. I do not recall, amongst 200-some graduate students, any AA student who was doing PhD, though there were 18 doctoral students from India alone. In 1973, through a friend working in a historically Black college in Mississippi, I was assured of a faculty position, but it did not materialize as the college found an AA. Well, that is how my destiny brought me to UNLV in 1974 for a one-year visiting assistant professorship - almost forty-six years ago.

Finally, a history of mathematics sidebar note: my Department of Mathematical Sciences, 63 years old, has yet to interview the first AA faculty member for a tenure-track position! On the other hand, the diversity of math faculty can be measured by the fact that out of 30 tenure-track faculty, only 5 are US born and the remaining 25 are foreign born (Brazil-1, Bulgaria-1, Canada-1, China-7, Cyprus-1, Haiti-1, India-3, Iran-3, Korea-1, Romania-1, Russia-1, Sri Lanka-2, Taiwan-2).

February 29, 2020

COMMENTS

Very interesting. I would not have done well in the quiz as the town I grew up in was a very white farm town in northern IL. Sadly, I didn't even know 1 black person until I went off to college. **George Buch**

Fascinating! Thank you. **Hal Whipple**

65. WHO IS A MATHEMATICIAN?

A mere look at this title generates all kinds of thoughts, norms and interpretations about who is 'qualified' to be called a mathematician?. Before going further, let me narrow the discussion down by adding that its immediate context is a course, **History of Mathematics** (Math 314) that I am currently teaching. In the midterm exam given last week, one of the questions was the following: ***Describe briefly the lives and contributions of three male mathematicians that have impressed you.*** (In balance, a similar question was on female mathematicians too.)

While grading the tests, a response to this question of the very first student was a bit unexpected. But that of the second student surprised me as he had included his elementary school teacher in his list of mathematicians! However, this element of surprise disappeared shortly as lots of students had listed their elementary, middle, or high school math teachers into their short lists. It is a combination of excellence in teaching and the caring attitude of their mathematics teachers that defined for these students what good mathematicians are supposed to be. In schools, any subject teacher has to be a bit of a counselor too, if a teen looks up to that teacher for any non-academic problem. Personally, in the culture and the era in which I grew up in the Bathinda (India) of the 1950s, I can recall numerous stories when my instructors influenced me beyond their disciplines. As a matter of fact, I excelled in courses taught by those overall caring instructors.

Getting back on track: the students in this course are all undergraduates. More specifically, out of 33 students, 26 are seniors and zero is the number of freshmen and 5, juniors. As far as their majors are concerned, the largest number of 16 is from computer science, only 2 from mathematics and 10 from Math Education (future school teachers). In the US school system, a math major is not licensed right away to teach in schools until a few education courses and one semester of student teaching are completed.

For a perspective on the math background, the prerequisite for this course is **Calculus II** (Math 182) - not including multivariable calculus. Therefore, most of these students may have only heard of doctorate in mathematics, but they cannot visualize that it entails publishable research in doctoral dissertations. By and large, a student's image of a math PhD is built on bachelor's and master's degrees. With that kind of rationalization, I accepted their answers as correct.

In almost 60 years of college teaching, I have seen, read and done/dealt with a lot of mathematics and mathematicians. My summing up of this question is in three C's, namely, Competence in the subject, Communication of the material, and Compassion for the struggling students - they define mathematicians. They may choose to work in any academic setting - from schools to hi-fi universities. However, it is the proportion of these three C's that needs to be adjusted smoothly at different levels in different institutions.

On the other hand, if any C is absent, then the holistic nature of mathematics is lost - without Competence, Mathematics is not nurtured - without Communication, Mathematics is not served - without Compassion, Mathematics is dehumanized. This is a distillation of mathematics in all phases - its creation, its transmission, and its utilization!

March 19, 2020 (Danville, CA)

COMMENTS

Very interesting. I liked that one of your C's is for compassion towards struggling students. I had never thought about that. **Drew**

While one may need to rethink to call them mathematicians, it certainly is an absolute truth that primary and secondary school teachers play a very very important role for developing not only interest in the subject they teach but also in motivating many of them to become subject experts. I still feel overwhelmingly grateful to my scheduled caste Primary School teacher of seventh grade (Vernacular final stage, having State Board exam then), Chotubhai Khodabhai Vasava, of my small village 'Kavitha' near Bharuch in Gujarat. So wonderful teacher in Math, Science and Gujarati language, he was taking even personal care during out of the school hours that not only I passed High School Scholarship exam in that year (to get some monthly stipend for all the four years during High School study till grade 11) but became so good in Math that in my old SSC class (grade 11, with State Board exam) of the town 'Palej', in 'The Palej High School', though 'Special Arithmetic' subject was not being taught, I was allowed to take that subject to be prepared on my own, and in which I scored highest marks compared to other subjects (81%) and I took math as my major during my UG study at MSU. So, indeed they are a motivating force for making mathematicians, even if they themselves may not be mathematicians as per accepted norms. I am sure many successful mathematicians must have had during their Primary and Secondary School level very good teachers. Regards, **JR Patadia**

Thank you for this one, Satish! This is a very nice culling of the properties of a mathematician, and I hope to remember to aspire to meet the requirements of all three in my daily work. I hope you and your family are well in these times, **Jake Blanton**

Nice job! Was I a mathematician in the third grade when I invented rational numbers? Am I a mathematician now? I enjoyed reading it. **Paul Aizley**

I love what your students responded to. Who is a singer? Not just someone who makes recordings! It's all just folks singing. Who is a mathematician? Not just ... (HoM)

66. (HISTORY OF MATHEMATICS) IN HISTORY

My approach towards teaching a History of Mathematics (HoM) course is to incorporate into it general history, humanities and religions as much as time permits. But this is now significantly curtailed due to the online instruction that has become instantaneously global in the last two weeks due to the Coronavirus. In online teaching, there is no way to veer off on a tangent, something that I normally do while explaining a point in a live presentation. It is a struggle to find my groove in this futuristic mode of instruction.

Anyways, while preparing brief notes from the Berlinghoff and Gouvea's textbook, the following opening sentence of a sub-section (on Page 35) stopped me right in the tracks. "Around the end of the 14th century, many different cultures around the world were producing interesting mathematics." That is a very sweeping yet catchy phrase that makes some readers pause and ponder over it. I took it as a challenge to actually survey the entire world so as to determine what had happened in the 14th century - all from the top of my head!

In the 14th century, North America, comprising the landmass known today as Canada and the USA, was populated by various native tribes. There being no navigational boats and ships capable enough to sail across the Atlantic, the Europeans had not touched the eastern shores of America yet. The surviving petroglyphs and petrographs shed little evidence on math, science and engineering, and there are no monuments and structures that have been discovered. Structures are the DNAs of STEM in particular.

To the best of my knowledge, this story is no different in the lands known today as Australia and New Zealand. In fact, major parts of Papua and New Guinea, northwest of Australia, are still living in the dark mathematically, and are much more primitive by any definition.

In the south of the US all the way to Peru, there are scattered archeological evidences that relatively small civilizations and prosperous nations existed in the 14th century - Inca, Aztec and Maya are some of them. The pyramids and stelae, the pillar carvings that the old cultures left corroborate remarkable achievements in math, metallurgy and mining. What a travesty of life that these civilizations were literally decimated by the Spaniards in the 16th century, but they are now being brought back to life by the progeny of the conquerors! A sidebar: over the last 15 years, I have undertaken six archeological trips in this region. Wikipedia has a lot to read on.

A question that often comes up in my mind is how come the natives of North America did not develop in STEM to that degree? Yet, an inverse question pertains to the present - how do you explain a gulf of difference in the STEM achievements between the countries in North America vs. countries in South America? I do have some perspectives and answers, but this is not the place and time for them.

Moving latitudinal on earth: the continent of Africa is still relatively in the dark and waiting to be excavated and explored mathematically. The exceptions are the massive pyramids and temples in Egypt - telling of its golden era 4000 years ago. Even Egypt did not have much for a while, but in the 14th century, Cairo became a center of Islamic theology and learning. It has one of the oldest universities, Al-Azhar, built around a famous mosque, and was established in 972 AD! There is every reason to believe that arts, sciences and commerce flourished in the Islamic nations of North Africa all the way to Algeria and beyond.

In the early Christian era, the Nordic nations were considered as being barbarians for lack of culture. Beginning in the 3rd century AD, the hordes of them came down south and gradually conquered parts of the Roman Empire in the Western Europe. The cultural assimilation continued. With the result, the western Europe including Russia was eventually homogenized culturally. The Church played a great role too. In contrast, in the last fifty years by accepting large numbers of non-assimilating Muslims from the Middle East, these nations are now dis-homogenized to the extent that they have already lost their historic identities! The 14th century was the beginning of great mathematics traditions in Italy and since then Western Europe has become a crucible of science and mathematics. After the fall of the Ottoman Empire, its influence extended into eastern Europe too.

In the 14th century, China had established the Yuan and Ming dynasties. Arts and STEM developed as mentioned in the textbook. In the 14th century,

Japan had borrowed some mathematical ideas from China. However, Japan has maintained its identity for the last two millennia and steadily developed in every facet of national life. Japan is the only Asian power that integrated modern science in its traditional way of life. There is quite a story behind it. In the first half of the 20th century, Japan, like some European countries, tried to colonize other neighboring nations militarily. But Japan was cut down to size after the World War II.

The entire Southeast Asia was a vast Hindu territory that gradually absorbed the tenets of Buddhism too. The art and architecture of the temples from the ones in Cambodia to those of Bali in Indonesia firmly established a prosperous region in the14th century. A poor person or nation does not build any structure! The evidence of mathematics in this region is embedded in the grandeur of their temples like in Angkor Wat of Cambodia.

A thumbnail history of present Indian subcontinent: before the 9th century AD, in the west upto Afghanistan, in south upto Sri Lanka and in east upto Myanmar, it was all a Hindu nation - considering Buddhism and Jainism as reformist movements in Hinduism. Afghanistan was conquered by the Arab Muslims in the 10th century, and gradually it became all Muslim. The attacks on Hindu India continued and all centers of Hindu learning were burnt and razed to the ground. Naturally, there was a great setback to the scholarly activities in north India throughout the 14th century. However, the Far East and south of India continued to work on and preserve mathematical traditions.

Well, that is my written spiel in analyzing a single sentence. It is like a popular phrase, 'It is obvious' in hardcore mathematics. Every math student falls victim to it sooner or later. Anyways, if any part of the world is left out (excepting Arctic and Antarctic!), one can apply the mathematical technique of interpolation! This was quite an educational exercise, which is worth playing with any century.

April 04, 2020

COMMENTS

It is often thought that the Vikings landed on the east coast of North America as some Viking artifacts have been found - I think in Labrador. **Francis**

Hi Satish:The 'reflection' with historical information. A couple of months ago, I saw a tv documentary about the existence of Hindu idols in South American countries such as Peru, Colombia, etc. Evidently, Southern India played a role. I am not tech savvy; I don't know whether I would have been ready to teach online courses had I not retired. The present crisis has given me impetus to let flow my creative juices, resulting in my freshly minted poem (attached). Yes, a single sentence can be a source for an extended discourse like a single cell turning out to be a full-bodied human or creative. Stay safe. **Moorty**

Bhai Sahib, you are quite a historian and a researcher. Enjoyed reading this beautiful piece of information. Thank you. **Sulekh Jain**

67. HISTORY-MATH-POLITICS

Good mathematics is neither done by mathematicians with empty stomachs nor in professional isolation. Social isolation is ok to some extent, as famously exemplified by Andrew Wiles, who proved Fermat's Last Theorem (1993), while working in total isolation at the Institute of Advanced Studies at Princeton. Yes, good songs have been written even in exacting conditions, which in fact make poets forget their dire conditions. Music too has that power over the body. People commonly tip or pay after listening to a street singer; whereas, no one would care about a person doing mathematics while sitting in a corner all by himself.

This train of thought took a hold over my mind as I was about to start summarizing a section on **Calculus and Applied Mathematics** in the textbook that is being used for the **History of Mathematics** course (Math 314) that I am currently teaching. Like every other textbook written almost anywhere in the world, the credit for developing calculus is attributed to Isaac Newton, the English mathematician and physicist (1642-1728) and Gottfried Leibniz (1646-1716), a German mathematician; in fact, a polymath.

Recently, George G. Joseph, formerly of Manchester University (UK), a Keralite, has written a popular book on the history of mathematics, ***The Crest of the Peacock: Non- European Roots of Mathematics*** (3rd edition, 2010). He has been able to dig new mathematical materials from the medieval treatises supposedly written in the Malayalam language of the present Indian state of Kerala. He has established beyond any reasonable doubt that at least two Indian mathematicians, Madhava (1340-1425) and Neelkantha (1444-1544) had the knowledge of infinite series and infinitesimals that Newton and Leibniz had used in the 17th century. In fact, nearly a dozen principles in physics, named after Newton, are derived using infinitesimals.

I have known about this calculus discovery by Indian mathematicians for 8-10 years. But it is one of the hundred such achievements of ancient Indians that are credited to Europeans and Americans - a long and sad saga. The following is an excellent link to a recent YouTube video by Rajiv Malhotra on this topic; https://www.youtube.com/watch?v=tR6QGe-a8gM.

Anyways, here are a few of my thoughts regarding calculus that have freshly sprouted out of my mind:

1. I do not subscribe to a theory that some Jesuit Christians in south India took these ideas from India to Europe. However, it may be added that Cyrian Christianity in India goes back in time to when the Apostle Thomas arrived (AD 52) on the Malabar coast of Kerala. He preached the Gospel and died in India (AD 72). There are all kinds of debates and controversies about him.

2. On the other hand, I do believe that identical ideas can shoot into the minds of intellectuals living thousands of miles apart. For instance, despite the ongoing controversy between the patrons and followers of Newton and Leibniz (never between the two directly), as to who brought out calculus first, the vast majority of math historians believe that they both did it independently of each other.

As a matter of fact the fundamental ideas behind calculus are absolutely innate to any contemplative person who is also curious about questions like, division by zero, infinity (potentially very large), infinitesimals (very small), infinite series, and mathematical limit, the grand-daddy of all behind calculus. There is clear evidence that Archimedes had an insight of calculus 2300 years ago, and so did the ancient Hindu sages/***rishis*** when they composed several mantras on the nature of infinity including the one ***mahamantra*** given at the very beginning of the **Ishopanishad** and **Kathopanishad**. Its English translation is: ***Om. That is infinite, this is infinite. From That infinite, this infinite comes. From That infinite, this infinite removed or added, Infinite remains Infinite. Om, Peace! Peace! Peace!*** Thus, calculus, crystallized by the end of the 19th century, and as now taught in schools and colleges, is the result of the collective effort of humankind of the last 2500 years. That is my latest stand on calculus.

3. However, I remain curious to know the political conditions in that part of India where these two mathematicians flourished, because it must have been conducive to their doing of mathematics. There is a reason to believe in it. Newton applied calculus in physics, Indian mathematicians used calculus in astronomy as well as in astrology. At the age of 80, I have come to another conclusion that what Einstein's Unified Field Theory attempts in the understanding of matter in the universe, so does the Vedic Astrology in

the understanding of the universe holistically. While Einstein's Theory is being worked on by mathematical physicists in many leading universities worldwide; tragically only the remnants of Vedic Astrology are seen - all is lost due to political conditions in India over the last 7-8 centuries.

4. Knowing the fact that geographical boundaries change on the average every 50 years, a quick Googling did not reveal anything on the social and political conditions of the region in which Madhawa and Neelkantha did their mathematical works (worth Googling) attributed to them. It is very easy to rule people after cutting them off from their heritage and history, and forcefully replace them by foreign culture and history of their victors and colonizers. That is what has happened to the Hindus in India since 1192. Consequently, the history coefficient of the Hindus has been reduced to zero (modulo a set of measure zero). One of my goals for the remaining years of my life is to rekindle the flame of history. History of Mathematics is my mode, tool and vehicle for it.

5. In 2003, I wrote a reflective piece, **Calculus Defines Civilization.** Therein, I have captured calculus in the west as a tiny stream starting in the 17th century. It turned into a mighty river by the 20th century - watering, fertilizing ideas, and unleashing various powers - including nuclear and colonizing ones. Currently, it is taking them into the colonization of the Moon and Mars! **My question is** as to why in India, this power of calculus was not harnessed in developing sciences.

Here is my answer: the stream of calculus in India had lost its way in an intellectual desert-scape. Not too far from the region where Madhawa and Neelkantha lived lies the ancient city of Madurai, a place of Hindu pilgrimage, in the state of Tamil Nadu. In 1335, Afghani Muslims ransacked the city, terrorized the Hindus by carrying out a genocide of untold proportions - turning it into an open graveyard for years. By the time these Muslims were defeated in 1378 by a Hindu king, the intellectual culture had disappeared from the entire region. That is the tragedy of calculus in India.

An intellectual person without the fighting spirit of a soldier is vulnerable - a perennial lesson of history.

April 22, 2020

COMMENTS

I think perhaps one of the reasons as to why Calculus never found any practical application in India was that it was used for Astrology rather than in the application of practical science. By the time Newton and Leibniz came along, Europe was already advancing towards the Industrial Revolution and so could see a practical use for this mathematical tool. **Francis**

I wrote: Thank you Francis for your comments and edits. In the west, way back in the 17th century, math and science were pursued in many countries including England, Italy, France, Spain, Germany, Russia and Sweden. Above all scholars were pursued by monarchs. About India, so little is known, surely calculus was limited in time and place.

Satish, I think it was Carl Jung that said great ideas spring up independently and the notion of the infinite and the infinitesimal, being great ideas, demonstrate just that. Take care and thanks for sharing. **Brian**

68. COVID-19 AND ONLINE MATH

Today, on a local TV channel, I just heard on the evening news that in the US schools, the progress in reading has declined or slided down by 30%, and in mathematics by 50% - both due to online instruction. Naturally, it struck a chord in my mind in this regard at UNLV. Live teaching was switched to remote or online instruction for the last six weeks of the spring semester. In fact, it was global - maybe the first time in world history when people reacted to an event particularly when it became a question of their survival.

So far, I am not aware of such a study done for college students. But I do not anticipate different findings, except in the category of reading skill which is not a key component of the curriculum at the college level. My opinions are based on a teaching experience of nearly six decades.

As a digressionary note, it made me think that the present-day US is driven by data no matter how biased it is, how skewed it is, how political it is, and how out of date it is. It seems that at present, more Americans believe in the power of data than in the power of God. No exaggeration! Of course, there is no study or data behind this assertion. In fairness, there are exceptions. In the class, while discussing any topic related to statistics, I place emphasis on knowing who is paying for the study and who is hired to do the study. Of course, the sample size, level of significance, and randomization of a sample are very pertinent to the conclusions. My favorite line is that you can always find a sample and level of significance to justify any statistical hypothesis.

Back to the heart of this reflection: the essence of mathematics lies in deductive thinking. Interestingly, that makes mathematics undemocratic - in the sense that a statement in the world of mathematics is not decided by any voting. Here, all mathematicians are always on one side. Reason: mathematics has no room for theories; it has theorems - all carved in stones.

Even at school level, mathematics is not just about spitting out solutions of problems by memorization or after practicing on a few examples. My experience of six weeks of online teaching is that I just could not impart to my students how to think deductively. Partly, there is no right instructional technology, and it is not easy to master the use of technology. There is one mindset that adopts technology, another mindset adapts to technology, yet in mathematics, few are adept in technology. Above all, the rate of change in technology is faster than that of typical faculty's ability to keep pace with new technology. Faculty engaged in research have only specific points of contact with technology. Thus, my online teaching was adversely affected during this phase.

On Day One, I tell my students about maintaining decorum of the classroom that is 'sacred' to me. This is thrown to the winds in online teaching. Last semester, in my Calculus III (Math 283) class, average attendance dropped from 75% to 25%. In the History of Math (Math 314) class, from nearly 100% to 60%. It affected my enthusiasm - as the empty chairs in an arena do to a performing artist.

In online teaching, a paramount feature of peer learning is missing. Through daily postings of the Office of Faculty Affairs, I have come to know of instructors and students having pets and other comforters around them. They might have been distractions to other students. If a student was emotionally affected by Coronavirus, then it was ok to observe consoling moments during the online period.

The bottom line is that the 2-dimensional online teaching is only better than the option of blanket cancellation of classes. No one has a crystal ball to tell as to when the old 3-dimensional normalcy will come back. At the same time, a fact of brain study or neurology is that in response to a new pattern of behavior, the brain starts rewiring the nodal points after 30 days. This will surely have far-reaching societal consequences.

June 25, 2020

COMMENTS

Dear Sir, Very significant opinion! Thank you very much for keeping me in the loop. With warmth and greetings, **NK Thakare**

Might be one of your best yet. And I couldn't agree more with the idea of making numbers fit the narrative. All we keep hearing about on the news is the number of positive cases...the denominator. No one talks about the deaths and how that rate is plummeting as the cases increase. Maybe since it doesn't fit their "control" narrative. **George Buch**

Many great points. Thank you for the email! **Drew**

69. RELIGION, MATH & POLITICAL POWER

What is Jain(a) Mathematics?, What is Islamic Mathematics?, What is Hindu Mathematics?, and so on - associating math with every major religion of the world. Does it mean that people of different faiths do math differently, or does math itself look different in different religions? As a humanistic historian of mathematics, my answer is an unqualified No. For instance, say, in basic math like, Quadratic formula and rules of calculus - they are applied in identical manner everywhere. The beauty of mathematics, though a man-made discipline, is that it is amazingly independent of nations, religions, cultures and gender.

Before proceeding on the subject, let me clear up the above reference to 'humanistic mathematics'. This is the study of the History of Mathematics (HoM) integrated with history in general, philosophy, political systems, and religions. Each one has a very strong connection with mathematics. The current piecemeal approach to HoM does not give the extensive picture of the development or decline in mathematical activities in a region or in a society.

In the context of X mathematics, one could say that in a certain time period, a particular body of mathematics was first developed by mathematicians of X faith, where X could be Islam, Judaism, Jainism, Hinduism, or Christianity, and so on. Or, one could even talk about the relative contributions in math and sciences of scholars from different faiths. This area is politically sensitive in HoM, but I raise it every time I teach a course in HoM both at the undergraduate and graduate levels at UNLV.

Here is a well known scenario in support of the above point. Anyone can see that in the last 100 -125 years, men and women of the Jewish faith have contributed far more in math and sciences than perhaps the contributions of scholars from all other religions combined, even though the worldwide population of the Jews is only 15 million. Whereas, in the same time period, the contributions of the Muslims with a worldwide population of 1.6 billion, is nearly zero. The 1.3 billion Hindus are not too far ahead of the Muslims. But the reasons for the low contributions of these two peoples are orthogonal. Anyway, the seminal intellectual contributions are measurable data; say, if viewed from the list of winners of the Nobel prizes, Fields medals, Abel prizes, and other international awards and recognitions.

This line of thinking was triggered three days ago after listening to 6-7 mathematical talks presented in an International Conference on **Jainism and Mathematics**. It was a 3-day (Dec. 12-14, 2020) event organized and hosted remotely by the **Jain Center of Greater Boston**. Due to the Coronavirus pandemic, this conference, scheduled to be held live in Boston last summer, was naturally moved online. Twenty-five papers, prepared by nearly 35 researchers, were read in 3-hour time slots (6 -9 AM Pacific US time) taking logistics into consideration for the participants from all major countries in the world. With the exception of 1-2 speakers, the rest were mainly from only 2-3 institutions in India. From their last names, it appears that they are all Jains - by and large, a tight knit community.

At the age of 80 plus, my curiosity is beyond the claims of scholars that great ancient Indian minds had discovered some results in math (like, Pascal triangle, geometry, transfinite numbers, zero, place-value of a number, large numbers, knowledge of astronomy and cosmology) long before the Europeans did. I want to find out the political reasons for this loss of knowledge and credits in India.

I have no qualms in saying that the oral and physical survival of several religious and non-religious treatises and traditions is due to the Brahmins in the Hindu caste system. The temples were used as depositories by long lineages of Brahmin priests - a positive side of the caste system. In a few temples in south India, the treasures have been protected for centuries, despite upheavals in the political landscape of the region. Seeing it is unbelievable!

A brief sidebar on the history of Jainism: Both Jainism and Buddhism are reformist movements of Hinduism. Siddharat/Gautam (563 - 480 BC), the founder of Buddhism and Mahavir (599 - 527 BC), the founder of Jainism were the crown princes of the neighboring kingdoms in north India (parts of present-day Nepal, eastern UP, Bihar and western Assam). Most likely, they were distant cousins too! Nonetheless, both of them renounced material wealth in favor of treasures of spirituality - thus, becoming far greater than the emperors. Sometimes, I wonder if they had not been royal scions; history may not have given this much credence to their achievements. Anyways, such are the paradoxes of name and fame in life. The worldwide population of Jains is less than **5 million** (Wikipedia), but their relative impact is significant.

The most prominent tenet of Jainism is non-violence, which is practiced by the Jain monks to a mathematical limit. I don't know of any Jain spiritual principle or specific meditation that would be conducive to the discovery of mathematical results or some secrets of life. After all, it is the focus and concentration of the mind that ultimately solves riddles and problems of a society. In Buddhism, ***Vipassana*** meditation does enhance overall sensitivity to a height that helps in the resolution of a problem. For the record, twice I have attended a 10-day ***Vipassana*** course (2007 and 2018).

On a personal note on Jains: while growing up in the Bathinda of the 1950s, the male and female monks were a common morning sight. Their bare needs were taken care of by the ***bania*** (a Hindu caste) community. Locally, all Jains were ***baniyas*** and vice versa. As a matter of fact, Mahatma Gandhi was born into a Gujarati ***bania*** family. In 1891, after his return from England, he came under the influence of a Jain guru, Ram Chandra. The principle of non-violence came to Gandhi from several directions. He eventually perfected it and empowered most Indians in their struggle for independence from the British.

Back to the conference: a talk on ***Karma Vibrations of Consciousness for Mathematical Estimation of Consciousness*** was of my type. The author's algorithm gives the 'consciousness coefficient' of an individual. Also, the talks on a more than one thousand-year old treatise, known as ***Siribhoovalaya*** was simply astounding in its scope and structure (worth Googling). It was written by Kumudenu, a Jain monk! Immediately, my mind started spinning on the following questions - applicable to all ancient scholars:

1. Who was Kumudenu - his date and place of birth, parentage, and early signs of his greatness?
2. Where did he get his education - gurukul/school and university - intellectual culture of the era?

3. What were the social and economic conditions of the region - including those of his family?
4. Most importantly, who was the ruler of the region in which Kumudenu spent his lifetime and what did he do in return for the royal patronage he may have received?

At the end of the presentations, a panel discussion was held as to how to bring out this ancient (Jain) knowledge in English, and in a form available to the West. It has to be understood that the Hindus started losing their political freedom in their homeland after the Afghani Muhammed Ghori defeated Prithviraj Chauhan in 1192. Eventually, every part of India, for a short or long time, was ruled by one of the hundreds of foreign invaders of India. They decimated Indian heritage, looted and razed temples to the ground, burnt down great libraries and universities, and carried out periodic ***katal -e-aam*** (total genocide) to keep the coming generations of Hindus terrorised forever.

That is why 80% of the present-day Hindus are still living in fear and have no fire in their belly to rule over their only homeland, India, or become a political force abroad, say in the USA. Political freedom is the most precious treasure that has to be defended every day! A lesson of history is that once a society loses its political freedom, then eventually it loses everything.

It is good to have intellectual validation from the West, but it should not be at the cost of awakening the Hindus in India. Even after 73 years of independence, there is no degree program in a college in India where a student can study sciences/math and Sanskrit at the same footing. Recycling superficial mathematical facts and figures is not seminal scholarship. Also, where is the incentive for anyone to study math and Sanskrit? Where is the employability of such graduates? Knowledge for the sake of higher knowledge has no institutional nurturing in India - public or private. Finally, it is ok to have part time scholars, as most of us are, but the great vaults of ancient scriptures cannot be opened with simple keys of rudimentary knowledge.

December 15, 2020

70. WHAT RAMANUJAN MEANS TO YOU?

This was the title of my keynote address that I delivered on Dec. 22, 2020 (10 AM, IST) at the one-day conference on the **History and Development of Mathematics**. This conference was remotely organized by the Mathematics Department of Maharshi Dayanand University, Rohtak, Haryana (est.1976). Mainly, it was a mathematical celebration of the 133rd birth anniversary of Ramanujan. In 2012, his birthday (Dec. 22, 1887) was declared as the National Mathematics Day by the then Prime Minister, Manmohan Singh. My guess is that Manmohan Singh's decision was also prompted by TP Srinivasan (1932 - 2013), his long-time friend and a distinguished mathematician. Incidentally, my Economics colleague and I (in Panjab University Evening College Shimla) had hosted a lunch for the Singhs and Srinivasan at our residence in November 1964 when they were holidaying in Shimla. This reflective article is extracted from this address and elaborated in a few places.

First of all, I explained that the **Humanistic History of Mathematics,** my current area of research, is the study of the History of Mathematics (HoM) integrated with history in general, political systems, philosophy and organized religions. Therefore, during the address, I interjected these disciplines at every opportunity. The very first moment came when I greeted the audience and told them of my addressing them from Las Vegas, the **Entertainment Capital of the World**. I refer to Las Vegas as the **Indrapastha** (the present region around old Delhi - not too far from Rohtak!) of the Mahabharata era (c. 800 BC). Thus, sciences and mathematics were highly developed in the India of the Mahabharata era as evidenced by the description of grandiose palaces, weapons of mass destruction, and entertainment centers etc. Mathematics is one of the indexes of affluence and power of a nation.

I pointed out that though this conference is about HoM too, but to the best of my knowledge, HoM is not taught in any regular course in India! This has to change - and the sooner the better. This absence is a reflection of the lack of interest in the general history especially amongst the Hindu masses and the intelligentsia alike (Hindus make 80% of India's population). It is no wonder that the leading nations in HoM are the ones that build economic empires and colonize the world, now space, with trade and the sale of armaments etc. - namely, the UK, France, Germany, Japan and the USA. The total autonomy of Indian colleges and universities (in the public and private sectors) alone can bring innovation in academic programs and in the production of seminal and indigenous research of the caliber of Nobel Prizes and Fields Medals.

What Ramanujan Means To You? is not a rhetorical question. I wanted the participants to ponder over it as to how and why Ramanujan's achievements have any relevance today. Otherwise, it is pointless to celebrate a life if it cannot be connected with the present. I also raised the questions: **Why has another Ramanujan not been born in India yet**? Is the Indian educational system flexible enough, say, to let an 18-year old get a PhD? I told the story of a 13-year old student currently enrolled at UNLV who already has four associate degrees from a community college!

The focus of my address was on my thesis, **On the Making of Ramanujan**. Its gist is that Ramanujan practiced ***Bhakti*** (devotional) ***yoga*** at a very deep level, and pursued mathematics with equanimity of mind - not at all concerned or competitive-who is doing what etc. Eventually, there came a point in his life when there was a confluence or a resonance effect between these two streams. It took place when Ramanujan's brain cells were over activated due to constant fever caused by the TB (Tuberculosis Bacteria). What now follows are five salient phases of his short life span of 32 years.

A. **Seeding of the Bhakti Yoga** (1887 - 1898)
Ramanujan's initiation into ***Bhakti yoga*** was subliminal - watching people daily, especially his mother worshiping the family goddess and singing in temples. He was the first born child. In December 1889, Ramanujan contracted smallpox, but he survived, unlike the 4,000 others who died in one district. For a child, it was natural to start believing that his survival was due to his mother's divine faith. In 1889, his mother gave birth to a son, but he died before completing a year. Imagine his mother's anguish and her becoming protective of Ramanujan. I can fully identify with this scenario as I witnessed my wife undergoing this experience in 1980 - her memory of that event remains vivid.

'Unfortunately', this happened to Ramanujan's mother twice again - in 1891 and 1894; a girl and a boy were born - both died within a year!! At such a point in life, people either lose their faith in their belief systems or become more devout. His mother remained steadfast. Ramanujan learned about the ***Bhakti*** traditions, worship rituals, and ***puranas*** (secondary Hindu scriptures). He was 11 years old when his mother gave birth to a boy in 1898, and yet another boy in 1905! These two boys not only survived the dreaded first year, but they outlived Ramanujan. His parents took the births of these two sons as divine blessings, and that faith cemented down in Ramanujan's life too. Subsequently, he always credited every deep mathematics theorem to his family goddess, **Namagiri Thayar** (Goddess Mahalakshmi) of **Namakkal**. He often said, "An equation for me has no meaning unless it expresses a thought of God. Great things in life are the results of yoga (***Bhakti yoga*** is one of the many) while physically engaged in any chore.

B. **Dips in Mathematics** (1897-1904)
The genius of Ramanujan in mathematics through the age of 11 is not known. There are no anecdotal stories - the reason being that he was raised in a small town and in Tamil culture that forbade any show-off. After the 1857 uprising in India, the British had iron fist control over Indians by enforcing punitive legislation that broke the will of the people. Total submission was the only way to survive. Such was the state of mind of the Hindus especially living in the eastern and southern parts of India. Thus the compliant regions were duly 'rewarded' - like, in 1857, universities were started in Calcutta (Kolkata), Bombay (Mumbai) and Madras (Chennai). For a relative perspective, the first university in Haryana state was established in Kurukshetra in 1957 - even ten years after India's independence!

Ramanujan's young mind was ignited when he got hold of SL Loney's classic book on **Trigonometry** and GS Carr's book containing thousands of theorems. Had Ramanujan not seen these English textbooks, his mathematics talents might not have received bursts of oxygen. He was so much fired up by mathematics that he neglected the study of other subjects. Consequently, he failed twice (1906, 1907) in the college FA exam. According to one source, he failed in math too as his solutions were beyond the understanding of his examiners! The reason he did not fail in high school was that the examiners there could follow his solutions!

C. **Wilderness Years** (1904 -1910)
Without the FA diploma, 20- year old Ramanujan had no steady job, but tutored math to survive. He lived in depression and dejection, and moved around. Such periods of soul-searching for one's identity and purpose in life are common traits amongst the geniuses. A common social practice in India was to get a disillusioned boy married with a belief that sex life (started with married life in that era - unlike today) would reset his life on a normal course. I know so many of my childhood friends in India who were married in their teen years. Siddhart (who became Buddha) was also married under similar circumstances. Ramanujan was 21 and his wife was 10 years old when they were married in 1909. The strength of a joint family system lies in its support system. For his spiritual affirmation, my guess is that Ramanujan may have met/heard of Raman Maharishi (1879- 1950), a great Tamilian yogi.

D. **Budding Mathematician in India** (1910 - 1914)
Around 1910, Ramanujan's mathematical prowess came to the attention of V. Ramaswamy Aiyer of Madras University, a founder of the Indian Mathematical Society (est. 1907). He showed Ramanujan's work to his English colleagues. One of the English professors sent Ramanujan's papers to mathematicians in the UK - Principle of Associativity at work. At this point, Ramanujan had already published some research papers before leaving India in 1914. Some English mathematicians did not believe that it was Ramanujan who did this work, and some did not understand his math - a common fate of a genius amongst the average - a human paradox. Anyway, Ramanujan was now out of depression. Hindus strongly give such a credit to the wife for bringing about a transformation in the life of her husband!

E. **Mathematical Explosions in Cambridge** (1914 - 1919) and **Fever Yoga in India** (1919 - 1920)
Before World War II, Cambridge University (UK) dominated by GH Hardy (1877 - 1947) and Gottingen University (Germany) dominated by David Hilbert (1862 - 1943) were the two great centers of mathematics in the world - at the same time the greatest political powers (mind the connection!) It goes to the credit of GH Hardy and the Cambridge University administration for admitting and supporting Ramanujan as an exceptional case. Hardy and Littlewood (1885 - 1977) polished Ramanujan mathematically as he

was a bit of a rough diamond until then. The speed of Ramanujan's ability to learn the foundations of mathematics cannot be explained except by the yogic power that Ramanujan harnessed subconsciously. Apart from problems that engaged Hardy and Littlewood, Ramanujan also produced great mathematics independently. His research productivity seems to have accelerated during his illness.

Ramanujan contracted TB within 2-3 years after his arrival in London/ Cambridge in 1914. As a matter of fact, TB was raging in the urban population of western Europe due to indiscriminate industrialization, neighborhood factories without proper ventilation, damp climate, no protective gears, and no understanding of TB. Anyway, mathematics including Mock Theta functions that Ramanujan produced in the last 2-3 years of his life remained beyond the understanding of the world class mathematicians for the next 50 years! For Ramanujan, mathematics was meditation and meditation was mathematics! He did not die, but rather, he flamed out mathematically.

Concluding Remarks

1. Ramanujan's full name is Srinivas Ramanujan Iyenger. The longer a person is remembered after their death, the shorter gets their names - like Confucius, Buddha, Alexander, Gandhi. Hindu nomenclature is based upon horoscopes. Iyenger is a Tamilian Brahmin caste. His parents must have chosen the name in consultation with their family priest. Srinivas(a) = Sri (God) + Nivas (lives, dwells) means the one who dwells in God. Ramanujan = Ram (God) + Anuj(an) (younger or dear) means one close or dear to God. Ram is not necessarily the hero of the epic of Ramanyana. Thus, Ramanujan lived up to his name! On a personal note, I try to live up to my name, Satish (means God of Truth or Gandhi's Truth is God! It is very challenging.

2. In order to demonstrate a contrast between math and sciences, I said that in science and technology, the principle, **Necessity is the Mother of Invention** works - like, finding a vaccine for the pandemic due to the Covid-19 virus. However, in pure mathematics, some results are proved way ahead of their applications, if any. Ramanujan's Mock Theta functions are such examples.

3. I explicitly posed the following questions: **Can we duplicate Ramanujan?** My answer is NO.

Can we identify, nurture, encourage someone to approximate Ramanujan ? Yes, to some extent.

Here is my way of doing it. At the end of the address, I announced a donation of Rs one lac (100,000) as the seed money for recognizing the top student in the traditional MSc (Math) program - starting from the previous five years, 2016 -2020. The award of Rs 5000 will be given in the name of my late friend, Manohar Lal Gogna, a self-made person, solid math professor and teacher, who had a big heart for his students, friends and relatives.

4. Finally, Before getting into the substance of my address, I told the audience that in order to make this address memorable for all of us, please email me or put in the chat your questions or comments after the talk. The top three questions will be awarded Rs 500 to 1,000 with a citation. No one submitted any questions! That speaks a lot!

Satish C. Bhatnagar
December 24, 2020

PS: A famous Ramanujan's anecdote is of his telling Hardy that 1,729, the number of the taxi in which Hardy came to see him in the hospital, was interesting in the sense that it is the smallest number that can be written as the sum of two cubes in two different manners. I told my audience, with due apology to Ramanujan, that my turning 81-year old last week was significant in a 'similar' way that 81 is the fourth power of the largest number that defines a human life span! Viva Ramanujan!

COMMENTS

This is a fantastic read! I thank you for these insights. In fact before I slept today, I told myself I will make it a point to call you more often. And here you are, with so much written already. Thank you. **Pradeep Kumar**

I enjoyed reading your article about Ramanujan. Reflecting on Ramanujan I wonder why India has not produced equally able scholars in different fields since India's independence 73 years ago? However this is not to overlook the fact that there are lots of India CEOs for companies such as Google and Microsoft and many more. **Vijay Kapoor**

Aha ! Professor Sahib, your presentation of Ramanujan is superb. I had read about him long back but the discovery of the संगम of "Bhakti Yoga" with his math brilliance shows your wisdom. It would not be exaggeration अतिशयोक्ति to say that during 19th Century Ramanujan had gone from Indraprastha (India) to England to enlighten it with his wisdom, so have you come from the same Indraprastha to enlighten the Indraprastha of America in 21 Century. Hence, I would like to say with all the emphasis at my command that: "Yes, another Ramanujan has incarnated in Indraprastha (India) and migrated to America to enlighten it." **Jadav**

You have done justice with the *WIZARD*. Your presentation covered almost all the facets of his *greatness*. I am very sure that the audience must have benefitted from your enlightening and illuminating exposition. Such *Thought- Provoking* exposures from an academician of your EXCELLENCE.......is no doubt a COMMENDABLE initiative. **Vinay Bhatnagar**

Dear Sir, I appreciate your act of sending this mail to me. It is informative, interesting, illuminating and intelligently presented. Thank you for your courtesy ! With warmth and greetings, **NK Thakare**

Good evening Sir! It is very nice Sir. We missed this, but a lot of thanks for providing. Regards **Dr. Gauree Shanker**

71. IGNITING HISTORY (OF MATH) IN INDIA!

Both the United States and India have nearly 6000 colleges and universities of all shapes and sizes. For a perspective, the geographical area of the US is approximately three times that of India, but it has one third of the population. However, there is one mathematical characteristic that differentiates the two groups apart. Every US college has at least one undergraduate course in the History of Mathematics (HoM), that is required for prospective school teachers. Its popularity may be measured from the fact that each time I teach this course, nearly 75% of the students are drawn to it from other majors. In contrast, to the best of my knowledge, such a course (called a paper) is not even offered in any institution of India! Come to think of it, it tells a whole lot about the two people and the two countries.

This story does not end with HoM, it is far more stark at the level of history in general. For instance, in the US, no one can get a bachelor's degree without having taken a course in history of the US or of its state. Besides general courses in history, there are topical history courses all over the landscape - like, history of arts, history of dance, history of engineering, history of medicine, history of economics, and you name it. In summation, history is in the air of the USA that you breathe in, you see it, you touch it, you smell it, you taste it - it is all around you. Amazing! It seems everyone in the US is out there to create history of one kind or another - at either extreme of life's aspects - thus, making the US both a land of opportunities and a land of over achievers. This realization dawned on me 10-12 years ago after having lived in the US since 1968.

Recently, a friend of mine said that history is the collective memory of a family, a society, or a nation. Think for a moment - what good is the life of an individual who has completely lost his/her memory due to an accident, disease, old age, or dementia. It is much easier to enslave people or colonize them, once the masters or the victors have decimated all the records and monuments of the slaves and the vanquished ones. The importance of history at any stage in life cannot be overstated.

This thinking re-engaged my mind when I hit upon my old communication in a file. A few years ago, I decided to do my bit for my native India about kindling interest in history through HoM. As a life member of the Indian Mathematical Society (IMS) since 1981 (yes, 40 years), in September 2013, I proposed to its officers about instituting two awards in the HoM - one pertaining to the era before 647 AD stretching all the way back to the Vedic period in the BC era, and the other for a specific period between 647 - 1192 AD. A word about the two years is very significant. The great Hindu Emperor Harshvardhan died in 647 AD and in 1192, the other Hindu Emperor, Prithviraj Chauhan lost the decisive battle, his head, and his empire to the Afghani invader, Mohammed Ghori.

The IMS took more than three years to institute these awards. However, an ongoing challenge lies in due publicity of the awards and in having a good pool of potential awardees. The current resources of the IMS are limited. In case one is wondering about my not pushing for a course in HoM in an Indian school or college, then that is out of the question. The curriculum of every academic program is still shackled as it was before India's independence in 1947. There is no freedom to study and pursue any discipline as one does in the US. The academic programs are so deeply structured that the university departments are like silos - without any cross fertilization of ideas in curriculum, faculty research or community outreach programs. Presently, state universities are stuck in inertia due to the large number of colleges affiliated to them. The private universities are simply money makers. However, looking at current changes, I foresee the education system in India turning a corner in the next 25 years. Political systems impact HoM too.

A word about the HoM awards with the IMS: the minimum age is 35 years - that is when a sense of history begins to develop. Incidentally, the Fields Medals, the highest awards in mathematics, are not awarded to mathematicians over the age of 40, as creativity for seminal work in mathematics starts declining after one is in one's thirties. In a way, HoM is on the opposite ends of a totem pole. I have been passionate about HoM for the last two decades, and I try to pass on my enthusiasm for it to my students.

As far as eligibility for these awards is concerned, a person must have a master's degree in any discipline with some evidence of research in any aspect of mathematics - like, the history of a mathematical topic, life of a mathematician and their connection with political and economical conditions of the region in which they were supported. There are no PhDs in HoM in any US university! A holistic or humanistic approach is encouraged. The material for the award consideration could be a book, article - published, submitted or presented at a meeting. It is all open and flexible in order to generate interest in HoM.

In conclusion, this effort in pushing for HoM in India is like sowing seeds in virgin soil. They are sure to sprout with caring hands.

January 13, 2021

COMMENTS

I fully agree with you. But, things won't change here, I know for certain. As earlier conveyed, most colleagues in all institutions and universities are not open, suffer from pride (even if not good researchers) and prejudices, complexes and self righteous egos, mutual rivalries, and all the typical Indian mindset. I don't have hope, and pray to God whole-heartedly that may in due course I be proved completely wrong!!
Regards, **J R Patadia**

History may be taught in US Universities, but the average American's knowledge of history and world affairs is dismal. **Rahul**

Interesting. I do wish the interest in history here in the US was as great as you suggest. Perhaps, compared to India it is. But it seems to be waning. And, I think it is in fact possible to get a BA here at UNLV without history (as long as one fulfills the Constitution requirement in political science or another field, which is possible). General Education requirements seem to pertain to broad categories like humanities and social sciences. History is a good way to meet the Humanities requirement, but not the only. **Paul Werth**

72. MY MATHEMATICAL DEBTS!

A few days ago, I explained to my wife of 58 years that all the material luxuries that our family has enjoyed and dreams that have been fulfilled are largely due to Mathematics. Since 1961 (yes, 60 years and going), teaching mathematics, researching mathematics and writing about mathematics have been my only source of bread and butter. As a matter of fact, mathematics has given me a lot more than this when it comes to my ability to comprehend life, as deductive thinking and compactness of mathematics adds a new dimension to a thought process and expression. Therefore, it is time to give back to math and in the name of math. I am now 81 years old, and this debt to Mathematics must be lightened to a certain degree. The best way is to popularize math, recognize people who impacted my life mathematically, and promote excellence in math with awards and recognitions etc.

This stream of thoughts erupted last night when I stumbled upon my three-year old correspondence with the Indian Mathematical Society (IMS) about my proposal of instituting three math awards for three top high school students in India. In a nutshell, this did not materialize. This morning, it struck me that instituting an award in a college or university separates the prevalent cultures of India and the USA.

The Indian mindset is still fundamentally shackled even after 73 years of independence that came to most of the Hindus (80% of India's population) after nearly seven centuries. Freedom to think collectively does not follow political freedom immediately. For example, it is due to the legacy of subjugation that people in India start protesting on the streets, if changes are made in any aspect of life- say, in education curriculum, laws enacted to curb black economy, removal of local tariffs so that trucks can run from one state to the other without stopping for hours to pay border taxes, granting Indian Citizenship to the non-Muslims persecuted in the Islamic nations of Afghanistan, Bangladesh and Pakistan; Kashmir integration, and so on.

The latest being the agitation against the Farmer bill, which was circulated as an ordinance for six months before it was passed by the Parliament. Interestingly, it is almost in the manifesto of every political party! The 90% of India's farmers are happy with the bill. But with the support of anti-Hindu global forces, the 10% of the farmers are aggressively running over the will of the 90%! To me, as an historian, such are the signs of the beginning of the end of democracy in the US and India. If you do not win in an election and in the Parliament, then you protest and put the public at inconvenience beyond limit. Ominously, the pendulum of democracy is shifting towards anarchy and dictatorship. The rule of law has become a sacrificial lamb.

Well, I don't want to digress from the subject of instituting the awards. The main point is that at UNLV, it has been so easy that in the last twenty years, we have set up ten Bhatnagar Awards for the top students in Arts, Dance, Education, English, Honors College, Mathematics, Music, and Theater. Nevertheless, I will continue to plant and transplant good ideas in India despite limited success so far. Pioneers have always faced struggles and resistance.

There being a bit of history in these communications - both local and global, I decided to extract relevant texts. No criticism is intended hee. After all, I was a part of that milieu before leaving India in 1968. But India has not left me!! That is a boundary condition of this reflection. The relevant portions from each email is given in the bold and italics below:

1. From my September 23, 2018 email to the IMS

In order to create and promote a nursery of future admirers of mathematics and increase the attendance at the annual IMS meetings, the following idea has bolted into my mind:

Recognize the top three students of the 12th class/grade in the prize year on the basis of an online mathematics competition. The students should be from the high schools situated within a radius of 11 miles from the site of the IMS meeting. The idea being that over a period the entire country would be covered for all intents and purposes. Also, this may generate a competition in hosting the IMS meetings. About 200 persons attending the meetings of the IMS, one of the oldest math organizations in the world, means it needs revamping. Involving promising students and their math teachers/coaches is one step in that direction.

It is not out of place to add that I am still fully employed at UNLV and intellectually active. However, I am cognizant of the looming shadow of

the mighty Time. With these broad boundary conditions, I want to donate Rs 510,000 for instituting three mathematics prizes in the amount of Rs X, 2X, 3X, where X depends upon the income generated in a given year by this endowment.

2. From my October 02, 2018 email to the IMS

The following are the important reasons to recognize the top three math students of the 12th class/grade of the host city, and involve them in the activities of the IMS:

Dispersion in India means 'good' students' move away from mathematics right after passing the 12th class. During my India visits, I invariably speak to the 10th graders and survey their professional interests and goals. There are hardly any boys or girls who want to be fundamental researchers or problem solvers of any kind. Almost everyone wants to be a physician, engineer, IT professional, IAS officer and commissioned officer in any military branch - wherever money and prestige are in plenty. Before such academically smart students do get away from mathematics, it is timely to plant a few mathematical seeds in their minds by recognizing their talent in mathematics. That is all. Later on, in their leadership roles in life, they are likely to support mathematics rather than oppose them.

Mathematics is for the young. Some of them may get turned on in the IMS meetings by learning the newest applications of mathematics. Mathematics is like manure to be spread in the fields - making every soil good for any crop.

Attendance at the IMS meeting should be viewed like going to a restaurant, where only a few diners are eating. Unless the business practices are changed, such restaurants would soon go out of business. Generally, the local organizers from the host institution, the IMS officers and some VIPs displaying corsages on their lapels make the vast majority of the attendees at the IMS meetings.

3. From the IMS email of October 02, 2018

I also agree with your view "Catch them young". But the question is How? and why? The present situation of employment in India for Math

students (other than IIT, IISc, TIFR and other similar institutes) is TOO bad. For example, in Maharashtra state, several teaching posts in colleges and Universities have been vacant for the last several years. The Govt. has put a ban on recruitment Next year, there will be elections and so Govt. may allow 20% posts to be filled !!! When these posts are opened there will be an auction!! The only choice for a math graduate at present is to start tutoring classes. He earns sufficient and for that he need not be a PhD.

The scenario at the UGC is very funny. People sitting in it count the number of papers, number of PhD produced, number of conferences attended/organized etc. that determines the grade of the institute. So many international journals are being published from every corner of the country. Conferences with "International" in name are being organized even at small places, where there is no participant even outside the district. Physicians or engineers may get something. These are my personal views. However, we shall take up the points raised by you in our meeting and IMS will try to do something to attract students towards mathematics.

4. From my October 14, 2018 email to the IMS

....I sensed a tinge of dejection and surrender to the present state of mathematics - in terms of jobs, PhD students, and employment conditions etc. Those of us who are 60+ had different experiences as students and faculty in India. Change was slow, bureaucratic, government controlled - no innovation! But look at the growth of colleges and universities in India in the last four years alone. Today, private universities may be money-making shops, but they will have to submit to the forces of the 'free market' in the next 10-15 years.

For instance, in 1960, the state of Haryana had only one deemed university in Kurukshetra; Punjab had none! (Assuming PU Chandigarh belonged to both Haryana and Punjab). As of today, my 'backward' home town of Bathinda alone has 4-5 universities! Math is everywhere - like, termite or a colonizer. Look at the number of college-bound students. In 1955, I was one in 1000 to attend a college. Today, the ratio may be one in three. The future of mathematics in India, for the next 20 years, is at

the interdisciplinary level. It will demand creative courses. During my India visit, I felt optimistic about mathematics.

At the IMS level, you think big collectively for mathematics, but act small for mathematics - a mantra. My emphasis is on collective wisdom. An individual may think big and act big, but soon he/she would fall big and be gone and feel disheartened.

The main point is that we all have enriched our lives professionally through mathematics. It is time to give back to mathematics in any shape and form. Recognizing three local students who have passed the 12th class exam is one small step, but it will go a long way. The selection is not to be at the national level at all. It is very local - students from the host city. Let the host local faculty and high school teachers decide on the top students. The IMS should just hand over the checks. Evaluate the award process for the next year, and make suggestions to the next host city. This recursive process will eventually strengthen the awards. The IMS should focus on the publicity of the awards and awardees!

5. From my November 20, 2018 email to the IMS

I assume that the Prize committee would be deciding on the institution of three new prizes for the 12- pass high school students in the region of the host institution, say, within a radius of 20 KM with a center at the IMS meeting site. ....The more I think about these prizes, the stogner I am convinced that such awards will popularize mathematics, raise its profile amongst high school students, which are the real nurseries of mathematics. Also, it will be good for the IMS as its outreach activity in the sense that its meetings are not limited to the university professors - projecting an elite image.

Again, I repeat that any IMS' logistic support to the host institution is the best way to identify the top students. At one time, I thought of some competition to choose the three top students, but that is not practical. The state and central school board results are the best sources for the initial data. All the host committee has to do is to check the top three scores in mathematics. In case of a tie, the scores in a common subject like Hindi or English may be used to break it. The prizes are intended

to recognise sheer excellence in mathematics. I look forward to hearing the decision of the committee.

6. From December 02, 2018 Internal email of the IMS bcced to me

As per the decision taken in the (adjourned) Council Meeting on November 29, 2018 I forward herewith the word file containing the e-mail communications taken between Prof. Satish Bhatnagar of University of Nevada, USA and certain members of the IMS Council with a request to go through the contents, study the proposal and provide your comments, input, suggestions and feedback in regard to the Proposal. Please feel free to provide whatever is felt in this regard.

7. From my December 07, 2018 email to the IMS

These awards and prizes are my way of paying off a part of my Guru Rinnh (Teacher Debt) and Rishi Rinnh (Heritage Debt). Squaring off three perennial debts, the Pitra Rinnh (patents debt) is reasonably paid off when grandchildren are ready for parenthood.

This moment reminds me of the era in India when the IMS was founded in 1907. It was the brainchild of a few British officers posted in India, and mainly British college professors, who were paid in India two to three times their salaries in Britain. They were intellectual colonizers. Besides, they wanted to stay professionally at par even while in India by pushing each other intellectually within the structure of a society. In fact, this is the history of all pre -1947 academic societies, colleges, universities and government infrastructures in India.

Let me make another historical connection. We all take pride in the mathematical achievements that took place in ancient India. No matter how much the political landscapes have changed in the Indian subcontinent, what has not changed is the Indian mindset - its love for philosophical concepts and abstractions, which are absolute prerequisites to the nature and development of seminal mathematics. Through these prizes, the IMS would seed the young minds in the spring of their lives. By following the lives of these prize winners for 15-20 years, the IMS may have unique data for statistical analysis.

Therefore, I need your support - in the approval of these prizes, in publicizing them, and in expanding the base of the endowment in future. Collectively, more is achievable. Thank you in advance for being a part of this mathematical venture.

8. From my January 02, 2019 email to the IMS

For the last few days the following story of a 16- year old US student, finishing 18 years of high school curriculum, and 4 years of college curriculum, has been making headlines all over the internet. Most of you would have read it too. This kind of scholastic achievement is not uncommon in the US. In 1968, the year I came to the US, one Michigan student was awarded a Math PhD at the age of 18!

"Braxton Moral was deemed "really, really gifted," from a young age. So gifted, it seems, that this spring the 16-year-old is expected to not only pick up his high school diploma, but a degree from Harvard, too. The Kansas teenager is set to graduate from Ulysses High School on May 19, 2019, then attend university ceremonies on May 30 to receive his Harvard undergraduate degree, he confirms to TIME" Yahoo Home page, 12/31/18.

I am fully aware that such stories have a long way to come out of the Indian education system due to many social, political and cultural constraints. But the IMS can take epsilon-delta steps by paying attention to the high school kids. That is the reasoning behind my proposal for high school awards.

The Y-2018 has already ended, but I haven't heard from any member directly. Hence, I thought of checking on the status of this proposal. Please let me know if I can help the Council in taking this momentous decision.

9. From January 02, 2019 email of an IMS member to me

Frankly, I indeed appreciate greatly your indulgence in pursuing colleagues to accept your offer, and such sustained efforts. I think no one needs to be convinced about the noble and necessary objectives/cause

behind your offer, every one must be agreeing. I also feel sure that many of our colleagues must be appreciating your offer and efforts, it is just that some could be busy and some taking for granted that the office bearers will surely accept the offer and some may be unresponsive just out of little habit of usually remaining unresponsive. As such, expressed explicitly or not, implicitly the colleague must be in agreement with what you convey.

I personally feel sure that the offer will not only be accepted but will be implemented too from the very next year - sans, perhaps, some/ little modus operandi of the process of implementation. At least making efforts to motivate young school children is indeed a small step in the right direction, we all feel.

10. From my March 12, 2019 email to am IMS member

After another two more months of wait, I thought of checking with the IMS Council on the status of my newly proposed prizes in mathematics. As a matter of fact, while sitting at the JMM meeting in Baltimore on January 18, 2019 I was pleased to see Ravi Jagadeesan, the winner of the 2019 Morgan Prize. He credited his math achievements to his parents and grandparents - catching the talent when young! My vision is similar - to identify and promote mathematics achievers right after high schools - not after college - too late for India.

Let me know, if I can be of help in its deliberations that may be going on about these prizes. I hope its logistics are clearly worked out so that the prizes are given out without fail. I have experienced that it is one thing to institute an award, but quite a different thing to ensure its disbursement is made to the rightful awardees.

March 15, 2021

PS February 03, 2021. I never heard from the IMS again on this subject. I too stopped pursuing it as my mind was engaged in other projects.

SECTIONS V

OTHER PERSPECTIVES

73. WHAT HOM MEANS TO YOU!

(Modified from Section V of Volume I)

Reflections in the first four sections give my varied perspectives on History of Mathematics (HoM). Like, at the time of Volume I, I again approached my colleagues and friends both in India and the US for their perspectives on HoM - especially from those who had spent a good time in mathematics and were personally known to me. I did not want to include or quote the perspective of any stranger irrespective of their reputation. Coincidentally, the number of essays received remained five as was for Volume I. I must add that the competitive US academic environment breeds weird collegiality and collaboration.

An underlying fact of life is that interest in some aspect of history is integral to the making of an intellectual mind. As we age, a sense of history is unavoidable. It eventually rubs into our social and political issues, and in particular, on our discipline of mathematics too. I have seen it happening to most mathematicians in their graying years. Distinguished mathematicians of the caliber of Abel prize winners are seen participating in the sessions of HoM at the annual joint math meetings (JMM).

There were no boundary conditions on these write-ups except that be in at least 200 words – no upper bound! In order to avoid duplication of scenarios, they may be gently edited. Any personal encounter with HOM was welcome. History of any mathematical topics of one's fascination and research are great. On the humanistic side, included are the lives of mathematicians, math teachers who may have touched and inspired you. It could be from your experience of having read or taught HoM – not excluded are math curriculum trends and teaching etc.

Mathematics and history are born to be apart - raised differently in a sense that for the first 15 years of life, math is treated like a privileged one, as child prodigies are found mostly in math, besides in music. History, for all intents and purposes, is not even born then. In high schools, math feeds on deductive thinking and problem solving. Whereas, history bogs down the young students in memorization of the inane dates of the life spans of history makers and historical events.

Different types of mindsets are required for each discipline. The intellectual development diverges in college. In the US curriculum, a math major does not take a history course beyond one or two required ones in the general education core. Likewise, a history major rarely takes a course beyond a rudimentary math course - again required in the general core. This fork widens up when graduate degrees are pursued.

However, a funny thing happens on the way to adulthood. While growing up, one cannot absorb math from the air. Math demands serious commitment. For example, for freshmen students, two hours of studying for every class hour is a must. On the other hand, history is associated with a mature mind. A saying goes: history is for the wise. That is why it is ludicrous to talk of child prodigies in history. Nevertheless, every adult absorbs history by reading books and magazines, debating and listening to talk show hosts on radio and TV. Essentially, you don't have to have a history degree to understand and participate in any aspect of history. On the other hand, math defies and frustrates any self-taught learners.

This note is an attempt to understand the non-intersecting nature of the curves describing math and history. GH Hardy (1877-1947) was the first mathematician who put it in black and white that creativity (Fields Medal type) in math starts sliding down after the age of 30. I liken such research mathematicians with heavyweight boxers. That is why mathematicians over the age of 40 are not even eligible for the Fields Medals. Ironically, this is the age of the coming of budding historians!

Because of the academic culture and institutional pressures, research mathematicians continue to spend their professional lives on a beaten path and hope for a 'Viagra' moment. It is in the lower tiers of institutions, where math professionals branch out into other areas once they are free from the pressures of getting tenure and promotions. This is amply supported by faculty participation in Chautauqua courses from small colleges and universities. Most of the time, I am the only one from a PhD granting institution. Fortunately, my learning curves of history and math have many points of intersection due to my bizarre path of education. In general, there is no blueprint for the making of a historian of mathematics.

In closing, I venture to remark that after 50 years of age, mathematicians are psychologically ready to branch out – be it in HoM, mathematics education, or philosophy of mathematics. Unfortunately, in the US, there are no graduate programs in HoM. Interestingly enough, some UK universities do have them.

Satish C. Bhatnagar
June 30, 2021

74. MY MATHEMATICAL JOURNEY & CORONAVIRUS

[Author's Note: George Buch, about 50 years old, is one of the half a dozen lecturers (Math PhD not required) in the Department. He has been known to me for almost ten years- first as a graduate student in my History of Mathematics course, and now as a colleague once he finished his MA with Mathematics Teaching concentration. Using his instruction expertise in flipping the classroom, he has attained the highest pass percentage in a precalculus course. George's ability to solve a pedagogical problem can be traced back to his working as a plumber for 20 years in his father's business. The mindset of solving a problem completely remains the same. After the age of forty, plumbing work starts taking a toll on the body. That is when George decided to become a teacher and came back to school after a gap of 20+ years. The way he reads my reflections, I am sure of his writing a popular math book one day.]

When Dr. Bhatnagar first asked me if I would be willing to write my perspective on the History of Mathematics (HOM) for his second volume of the ***Darts on History of Mathematics***, I was honored. First, because of all the faculty I have worked with, I am not sure I have encountered anyone as passionate about HOM as Dr. Bhatnagar. Second, as I sit in my home office amid this coronavirus scare, I feel we are in the midst of a serious time in which mathematics education, or the absence of it, will impact our near future.

I first met Dr. Bhatnagar when I returned to the University of Nevada, Las Vegas, to complete my master's degree in mathematics with an emphasis on Teaching Math. Dr. Bhatnagar created the program and we became friends almost instantly. I was continually picking his brain for things he had found worked (or did not work) over his teaching career. Also, during this time, I was introduced to the idea of the "flipped classroom". This was the concept that said instead of teaching in the classroom and having the students struggle on their own to complete assignments, instead, the students would watch pre-recorded lecture videos and we would tackle these tough homework problems together. For me, this marked a critical point in my math history as it was making use of technology not previously available to students like myself as an undergraduate student.

Little did I know that only a few short years after implementing those changes not just in my classes but by many within the department, I would be faced with an even more challenging obstacle, the coronavirus. In March 2020, our world greatly changed as everything began to shut down in fear of massive worldwide death from this unknown virus. Our campus closed and the remainder of the semester was to be taught "remotely" through online video conferencing applications.

Despite the fact the death rate never got close to the predicted levels (a separate discussion in the HOM), out of an abundance of caution, the following semester was also to be taught remotely. I mentioned earlier my concern for the lack of math education. For as valuable a tool technology had become for my teaching methodology, it had now become just as much a hindrance. The greatest aspect of the flipped classroom was my ability to see and help the students work the problems thus reinforcing the learning process. But now, as I was forced to simply sit behind a computer screen and never actually get to help them as I would like, they turned to resources like Mathway and Symbolab which would not only give the students the answer but provide the "work" for them as well.

It was this work that has led to my concerns. Remote students were performing considerably better on exams (20-25% better) than in-person classes. When asked to show their work, some students (15-20%) would provide steps that had never once been discussed in class. Now, I suppose they all had teachers in the past that showed them this EXACT same methodology, but that would be quite the coincidence.

As with everything in life, we learn and evolve, and these occurrences have forced me to do that. I would hope that my changes will ultimately have positive effects on students' learning. I chose math education as my career because there is nothing more satisfying than seeing a student that once struggled with math finally have that "aha moment" where they understand a concept and can now excel at it. There is no doubt in my mind that when future generations look back on the HOM, this will be a time that has a significant impact. I hope it will not be viewed as a time when technology set us back mathematically.

George Buch
December 2020

75. GYMNASTICS WITH HISTORY OF MATHEMATICS

[Author's Note: Petros Hadjicostas, a native of Cyprus, has been known to me for only three years since he joined the Department as an assistant professor. However, I have had far more exchange of ideas with him than with some of my colleagues known for thirty years. It is all due to the versatility of his intellect and engaging social personality. Not only is he well informed, but has an investigative mind. In the context of this book, it is worth adding that Petros resigned from his position as a tenured associate professor at Texas Tech University and joined Victoria University of Wellington in New Zealand. After working there for five years as a senior lecturer, he has returned to the USA. Petros is a rare breed of pure scholars who only care about their work environment - much less about their rank. In the long run, such faculty members define the academic culture of a university.]

Whenever I read about a result in Mathematics or Statistics that is named after a person, I try to find more information about that person and, if possible, find the original paper or book where the result first appeared. This is one of the ways I learn about the History of Mathematics and Statistics. In the past, when the internet did not exist (or was at its infancy), this was not always an easy thing to do. But today, with all the internet resources available, this is quite easy. One can also translate foreign language words and sentences using online translation websites.

In the second half of the 19th century and the first third of the 20th century, most of the top Mathematics and Physics papers (but not the top Statistics papers) were written in French and German. Many of these papers have now been scanned and they are available through the websites of libraries and foundations in various countries. The same is true for some old mathematics books. One may search the internet and try to find these "original" papers and books because they constitute a primary source of the modern History of Mathematics.

Next time one reads about a named result in Mathematics or Statistics, he or she should try to find the original paper where apparently this result first appeared. They might be surprised, and they might learn some interesting things they did not know before!

According to “Stigler’s Law of Eponymy” (which I am officially renaming “Hadjicostas’ Law of Eponymy” from now on), “no discovery or invention is named after its first discovery”. See, for example, Stigler (1983) even though it may not be the first reference that contains it! Thus, it is imperative to search multiple sources because a named mathematical or statistical result may have appeared in some similar form long before the “original” research paper or book where it “first” appeared.

Faà di Bruno’s (1855) formula for the chain rule of higher derivatives of functions was apparently known more than 50 years earlier. It appeared in a calculus book by Arbogast (1800). I actually did download the 2-page paper by Faà di Bruno from the web more than a decade ago, but the paper had no proof. (I needed the formula because I wanted to try and prove the infinitely many inequalities involved in Li’s (1997) criterion for the Riemann hypothesis. Of course, like many others who attempted to prove Li’s inequalities in the hope of proving the Riemann hypothesis, I did not succeed!) Faà di Bruno’s formula and various proofs of it appear in many other modern research papers and books. See, for example, Johnson (2002).

In a recent research paper, part of which I presented in a conference, I needed to use a Ramanujan sum, which is the sum of the m-th powers of the n-th primitive roots of unity. Introduced by Ramanujan (1918), it is well-known that it equals von Sterneck’s (1902) arithmetic function. (There are a few papers by von Sterneck related to this function or variations of this function, and I found all of them, but it will be too much to put all the references here. In addition, to fully understand what he was doing, one needs to look at most of these papers taken together. They are not difficult to read, but they are in German.)

Even though a Ramanujan sum equals a value of the von Sterneck function, it is not unfair to name the sum of the m-th powers of the n-th primitive roots of unity after Ramanujan because von Sterneck did not consider trigonometric polynomials like Ramanujan (1918). But here we have one more occurrence of Stigler’s law of eponymy! It was actually Kluyver (1906) who first introduced Ramanujan sums, more than a decade before

Ramanujan, and he proved important properties about them! I did find his 1906 paper and verified that! See also van der Kamp (2013).

Because universities are the main places where research is done, it is possible that one may work in a university where famous mathematicians or statisticians taught and did research in the past. From 2011-2016, I was working at Victoria University of Wellington in New Zealand. The front part of the main office of the School of Mathematics and Statistics, where the administrative assistants were working, had several paper models of three-dimensional polytopes (i.e., polyhedra) that were constructed several decades ago by the Scottish mathematician Duncan Sommerville. (He was actually born in Rajasthan, India, where his missionary father established a hospital.) After coming to New Zealand from Scotland, he worked at that university from 1915 until his death in 1934. He was a preeminent geometer, who wrote a number of research papers and books. (I have two of his books, including his famous book, ***An introduction to geometry in n dimensions.***) Until a few months before I left that university, Geometry was a course taught in that department by an older colleague of mine. Unfortunately, he passed away in early 2016, so the other mathematicians there decided to cancel the course he taught. I strongly protested because that was the place where Sommerville taught and did research in Geometry, but they ignored me.

Besides his books, Sommerville is known for the *Dehn-Sommerville relations* of a simplicial polytope, i.e., a polytope whose facets are simplices. (In two dimensions a simplex is a triangle, in three dimensions it is a tetrahedron, in four dimensions it is a pentachoron, and so on.) These relations, named after Dehn (1905) and Sommerville (1927), give equations for the number of faces of different dimensions of a simplicial polytope. For example, in three dimensions, if f_0, f_1, and f_2 are the numbers of vertices, edges, and planar faces, respectively, of a polyhedron with triangular faces, then we get $f_0 - f_1 + f_2 = 2$ (the famous Euler formula for planar graphs) and $2f_1 = 3f_2$ (two times the number of edges equals three times the number of planar faces, which are two different ways of counting the elements of the set {(v, e): v is a vertex of the polyhedron not adjacent to edge e such that v and e belong to the same triangular face}).

Actually, just by coincidence, I ran into one of Sommerville's papers (written in 1909), when I was working on the number of different kinds of cyclic compositions of a positive integer. A composition of a positive integer n is an ordered partition of n (e.g., 1+1+2 = 1+2+1 = 2+1+1 = 4 are three different compositions of 4). A cyclic composition of n is the number of equivalence classes of n, where two compositions are equivalent if and only if one can be obtained from the other by a cyclic shift (e.g., the compositions of 4 in the above example are equivalent). Inspired by his 1909 paper, I wrote at least three research papers on this topic. Sommerville (1909) presents a cyclic composition as a periodic sequence. For example, the above equivalence class {(1,1,2), (1,2,1), (2,1,1)} of a cyclic composition of 4 can be represented as the periodic sequence …, 1, 1, 2, 1, 1, 2, 1, 1, 2, … The paper is difficult to read (and quite long), but I look forward to completing my reading of the whole paper one day. (On the way, of course, I will write some more research papers on this topic!)

According to O'Connor and Robertson (1997), "In 1919, when the professor of mathematics at Otago University [in the south island of New Zealand] suffered a nervous breakdown, a young student there, A. C. Aitken, was left without support and Somerville [in Wellington, in the north island of New Zealand] began to tutor Aitken with a weekly correspondence." Had it not been for Sommerville tutoring young A. C. Aitken, I would not have been able to teach Applied Regression or Multivariate Statistics in various universities around the world!

It is said that Aitken was the first one (modulo Stiegler's Law of Eponymy) to use matrices in Multivariate Statistics and in Regression Analysis. His landmark article that probably changed the way Statistics is done when we analyze several variables at the same time is Aitken (1936). In this article, he introduces his famous estimator for ***generalized least squares***. His famous estimator is cited more often in Econometrics books rather than in Statistics books. In Statistics books, his name is often omitted! (Yes, the first time I encountered ***Aitken's estimator*** was in an Econometrics book even though I spent five years in a Statistics department getting an MS and a PhD in Statistics.)

Apparently, in another application of the universal Stiegler's Law of Eponymy, Aitken and Silverstone (1942) were the first to introduce (in some form) the famous *Cramér-Rao lower bound* on the variance of unbiased estimators of a parameter. See also Shenton (1970). The first sentence of

Aitken and Silverstone is as follows: "The present paper communicates some of the results given in a thesis submitted by Mr Silverstone to the University of Edinburgh in May 1939." Thus, if there is any dispute, A. C. Aitken and his student H. Silverstone knew about the Cramér-Rao lower bound at least since 1939!

Unfortunately, he took part in World War I with the New Zealand Expeditionary Force. He was wounded and his experiences in the war affected his mental health for the rest of his life. He died in 1967. See O'Connor and Robertson (2003).

References

A.C. Aitken (1936), On least-squares and linear combinations of observations, *Proceedings of the Royal Society of Edinburgh*, 55, 42--48.

A. C. Aitken and H. Silverstone (1942), On the estimation of statistical parameters, *Proceedings of the Royal Society of Edinburgh Section A*, 61(2), 186—194.

L. F. A. Arbogast (1800), *Du calcul des derivations [On the calculus of derivatives]*, Strasbourg, France: Levrault.

M. Dehn (1905), Die Eulersche Formel in Zusammenhang mit dem Inhalt in der nichteuklidischen Geometrie, *Mathematische Annalen*, 61, 561--586.

F. Faà di Bruno (1855), Sullo sviluppo delle funzioni [On the development of the functions], *Annali di Scienze Matematiche e Fisiche,* 6, 479--480.

W. P. Johnson (2002), The curious history of Faà di Bruno's formula, *The American Mathematical Monthly*, 109(3), 217--234.

J. C. Kluyver (1906), Some formulae concerning the integers less than n and prime to n, in: *KNAW, Proceedings, 9 I, 1906, Amsterdam*, pp. 408--414.

X.-J. Li (1997), The positivity of a sequence of numbers and the Riemann hypothesis, *Journal of Number Theory*, 65(2), 325--333.

J. J. O'Connor and E. F. Robertson (1997), Duncan MacLaren Young Sommerville, *MacTutor History of Mathematics Archive*, School of Mathematics and Statistics, University of St. Andrews, Scotland; available online at https://mathshistory.st-andrews.ac.uk/Biographies/Sommerville/.

J. J. O'Connor and E. F. Robertson (2003), Alexander Craig Aitken, *MacTutor History of Mathematics Archive*, School of Mathematics and Statistics, University of St. Andrews, Scotland; available online at https://mathshistory.st-andrews.ac.uk/Biographies/Aitken/.

S. Ramanujan (1918), On certain trigonometric sums and their applications in the theory of numbers, *Transactions of the Cambridge Philosophical Society*, 22(13), 259--276.

L. R. Shenton (1970), Letter to the editor: "The so-called Cramer–Rao inequality", *The American Statistician,* 24(2), 36.

D. M. Y. Sommerville (1909), On certain periodic properties of cyclic compositions of numbers, *Proceedings of the London Mathematical Society*, S2-7(1), 263--313.

D. M. Y. Sommerville (1927), The relations connecting the angle sums and volume of a polytope in space of n dimensions, Proceedings of the Royal Society, Series A, 115, 103--119.

S. M. Stigler (1983), Who discovered Bayes's theorem, *The American Statistician*, 37(4), 290--296.

P. H. van der Kamp (2013), On the Fourier transform of the greatest common divisor, *Integers*, 13, Article #A24.

R. D. von Sterneck (1902), Ein Analogon zur additiven Zahlentheorie, *Sitzungsber. Akad. Wiss. Sapientiae Math.-Naturwiss. Kl.*, 111, 1567--1601 (Abt. IIa).

Petros Hadjicostas
January 21, 2021

76. HISTORY OF MATHEMATICS IN INDIA

[Author's Note: Mahavir Vasavada, resident and native of India, has been a retired math professor of Sardar Patel University for more than two decades. Retirementent has given him ample time to involve himself in the activities of Gujarat Mathematical Circle, particularly of Prof. A.R.Rao Foundation, of which he is a convener. One of his favorite activities is to publish a fortnightly newsletter, where he reports about mathematicians and mathematical events of general interest. His wife, also a retired college mathematics professor, devotes her time in popularizing mathematical models as teaching aids in schools. Having known them for nearly 15 years has enriched me professionally.]

In most of the Indian Universities, the History of Mathematics (HoM) does not find a place in undergraduate or postgraduate curriculum. This is partly because of the general belief that HoM is about mathematics, and that is not a core mathematics. While there is some truth in this viewpoint, I do not see it as a reason to exclude the HoM completely from every mathematics curriculum. Even though the HoM is not mathematics, it is built around mathematics and mathematicians. On one hand, the HoM offers plenty of scope to expose a large amount of good mathematics, and on the other, it provides an opportunity to the learner to have a better perspective of mathematics which he already knows.

Mathematics is not created in vacuum, just sitting in an armchair. It is the outcome of the intellectual activity of a very high order, pursued vigorously for days and weeks, sometimes for months and years. Mathematics, once born, lives forever. Not only does a mathematical idea not die, it does not get outdated. For example, Euclidean geometry is taught in schools more than twenty-two thousand years after Euclid wrote it. Fermat's famous conjecture was based on his reading about Pythagorean triples which were at least two thousand years old at that time.

It is no small wonder that mathematical ideas are useful to mathematicians even after centuries, but it is a greater wonder that mathematical ideas are useful in other fields as well. Eugene Wigner [1] was so intrigued by this fact, that he talked about '**Unreasonable Effectiveness of Mathematics in the Natural Sciences**'. Wigner had written his article in 1960. Sixty years later, the effectiveness of mathematics has spread far beyond natural sciences. Prof. L. C. Young [2] in his book, ***Mathematicians and Their Times,*** writes: 'The ideas of mathematics after many centuries, can be incomparably more useful than laws laid down by the most successful of rulers, backed by overwhelming force.' That is a bite of the HoM.

If mathematics is not merely a store of knowledge, but is a creative activity interacting with human civilization all the time, would it not be worthwhile to know what mathematics was developed, when, why and by whom? The HoM tells these stories and more. It also tells us that mathematicians are humans, with human qualities, living in a human world. They have their failures and successes and when success comes, it does not come easily. So the HoM may have a lesson or two for our own life.

It is this view of the HoM that prompted me to teach a course in HoM to postgraduate students. While such a course would help every student, in broadening his outlook for mathematics, I had particularly those students in mind who would go in for teaching in schools or colleges afterwards. Through the teaching of history, I wished to convey to them the thrill of mathematics which they would then pass on to their students. To achieve this objective it was desirable not to teach history by presenting facts in a chronological order, but by putting it in a story form, going back and forth in time.

I tried to convey the beauty of mathematics, its surprises, and the continuity of its development. I told the students that mathematical ideas do not lose their relevance over a period and mathematics, though divided into branches, is essentially one monolithic discipline. I narrated stories of tragedy, struggle, humility, deceit, determination, and disappointment – the stories of Abel and Galois, Ramanujan and Hardy, Newton and Leibnitz, Tartaglia and Cardan, Bolyai and Lobachevski, Cantor and Frege. And, I talked about some serious mathematics in the course: Axiomatic approach, Non-Euclidean geometry, Indian contribution to mathematics, paradoxes of Zeno and Russell, Prime numbers and construction of regular polygons, and a lot more

Many students who went in for teaching commented that they found the course helpful in making their teaching interesting and learning for students enjoyable.

REFERENCES

1. Eugene Wigner: "The Unreasonable Effectiveness of Mathematics in the Natural Sciences," Communications in Pure and Applied Mathematics Vol.13 (Feb. 1960).

2. L. C. Young: *Mathematicians and Their Times*, North Holland, 1981.

Mahavir H. Vasavada
January 27, 2021

77. A GLIMPSE OF SOVIET MATH (PART I)

[Author's Note: Viktoria Savatarova and Aleksei Talonov are a married couple, in their fifties, both applied mathematicians, and natives of Russia (Russian Federation, a reincarnation of the USSR - acronym for the Union of Soviet Socialist Republics). I have known them for six years as my colleagues in the Department. However, their collegiality and willingness to support or collaborate in my projects is praiseworthy. Both of them are instructors and researchers of par excellence. A rare aspect of their personality is that both of them held high academic positions in Russia; however, in the US, they had no qualms in accepting jobs at the lowest rung of a professional ladder. Nevertheless, they are destined to rise in ranks like cream does in milk.]

Aleksei and I were raised in the country where secondary school textbooks on Mathematics and Physics were written by not only educators but also by experts who had made valuable contributions to their fields as researchers. For instance, our secondary school textbook on algebra was written by Andrey Kolmogorov, who had made significant contributions to probability theory, topology, classical mechanics etc. This book included the elements of differential and integral calculus.

School children start studying Algebra, Geometry and Physics in middle schools and continue it in high schools. Five years of algebra would guarantee certain skills and confidence in performing algebraic operations. Geometry was started in the 6th grade, and from the very beginning, students were supposed to learn how to prove theorems. Physics was introduced in the 6th and 7th grade, and then in the 8th through the 10th grade. It was revisited at a higher level - an effective approach. In this approach, a concept was first introduced to the middle school, where it served as a foundation for the next stage. When students became more mature, the ideas could be revisited on a higher level. As I mentioned before, derivatives and integrals were a part of the course of the 10th grade algebra. At that time, the requirement was ten years of schooling.

Sciences were very popular with students, and by sciences we mostly meant physics and mathematics. They came together, and were followed by chemistry, and then biology. Some schools specialized in physics and mathematics. Top universities ran evening or weekend classes for high school students. There were boarding schools too. For example, the boarding school at the Moscow State University was established by Andrey Kolmogorov, Isaak Kikoin and Mstislav Keldysh. On a personal note, we did not attend a specialist school, but got lucky to study at the universities which were considered as top schools in Mathematics and Physics in Russia.

I got my degree in Applied Mathematics and Physics from the well-known "Phystech" or Moscow Institute of Physics and Technology (MIPT). The school is located in a suburb of Moscow, Russia. The MIPT is a unique university in Russia. It was established after World War II by Peter Kapitsa, who won the Nobel Prize in Physics. Kapitsa researched with Ernest Rutherford in the Cavendish Laboratory at the University of Cambridge. After returning to his home country, he helped in establishing a research university on the model of Cambridge University.

The following is a summary of the key principles of the Phystech System, as outlined by Kapitsa in his 1946 letter arguing for the founding of MIPT. There was a rigorous selection of gifted and creative young individuals. In a creative environment, it involved bringing leading scientists in close contact with students. An individualized approach was to encourage the cultivation of students' creative drive and avoid overloading them with unnecessary subjects. The rote learning common in other schools, as necessitated by mass education, was bypassed. The idea was to impart education in an atmosphere of research and creative engineering while using the best existing laboratories in the country.

In its implementation, the Phystech system combines highly competitive admissions, extensive fundamental education in mathematics, as well as theoretical and experimental physics in undergraduate years. There was immersion in research work at leading research institutions of the Russian Academy of Science starting as early as the second or third year. It was before Russia started awarding bachelor's and master's degrees. We studied for six years, and at least the last two years were supposed to not only take classes but also conduct research at one of the research institutes of the Russian Academy of Sciences. I did my research at the Theoretical Department of the High Temperature Institute of the Russian Academy of Sciences. My first scientific paper was on mathematical modeling of chaotic behavior of charged particles (plasma). Mathematically speaking, I solved nonlinear differential equations. The catch was that the behavior of the charged particles that seemed to be chaotic was totally predictable!

After graduation from the university each of us continued as graduate students. Aleksei completed his degree at his Alma mater. I also got my "Kandidat Nauk" degree from MEPhI. A degree of a "Kandidat Nauk" is the Russian equivalent to the research doctorate (PhD) in other countries. The next degree for both of us was a degree of a "Doktor Nauk" which to some extent can be considered similar to the Habilitation in Germany. A "Doktor Nauk" degree is conferred by a national government agency called the Higher Attestation Commission. The candidate must conduct independent research. Therefore, it requires no academic supervisor. Usually, the candidate will be an established scholar. "Doktor Nauk" degree is conferred for a significant contribution to science based on a public defense of a thesis, monograph, or of a set of outstanding publications in peer-reviewed journals.

In the former USSR, this degree was considered a sufficient credential for tenured full professorship at any institution of higher education.

Viktoria Savatorova
March 04, 2021

78. A GLIMPSE OF SOVIET MATH (PART II)

Sometimes, it is very important to ask yourself a question: why did you do exactly this way? Of course, the question about your career is very important, and I often come back to the question, why applied mathematics became my major area of studies. I may be able to pinpoint in my mind the time when I chose applied mathematics over other majors.

I was an eleven-year old student in Moscow middle school in the Soviet Union, a country that does not exist any more. All students studied algebra and geometry during all seven years of middle and high schools. In the early 1970s, students used math textbooks written by Andrey Kolmogorov (1903-87), who made significant contributions in varied fields of mathematics - namely, probability theory, topology, classical mechanics etc. In the last year of high school, this program included elements of differential and integral calculus. The new math curriculum was part of the education reforms, which were intended to bridge the gap between school curriculum and the practical needs of the state, and to prepare students for entering the workforce upon graduation. The authors of the new curriculum were predominantly professional mathematicians and university professors.

My interest in mathematics was supported by very interesting articles in the journal ***Kvant*** (Quantum). Kolmogorov had been one of the authors and editors of this journal since 1970. ***Kvant*** published miscellaneous mathematical problems, puzzles, discussion topics, and interesting facts from the history of mathematics. A separate section was dedicated to study tips and sample problems for prospective university and college students. Also, ***Kvant*** published information about mathematical competitions, lectures and events for college students, in particular, in Moscow Youth Center "Vorobyovy Gory". The Vorobyovy Gory is a very beautiful place in Moscow. Besides school education in the USSR, there existed a whole network of special clubs that pupils could attend for free and study a wide number of subjects. All students of elementary, middle and high schools had opportunities to join hundreds of different groups free of charge.

I had a dream to join a mathematical or astronomy group, but unfortunately all places were occupied, and I had to choose the astrophysics group.

The students and postgraduate students from Moscow State University delivered lectures and discussion classes for us several times per week. It was a very unusual form of education and I was really surprised how the solutions of Newton's equations could be used for describing evolution not only of our solar system, but of galaxies as well. Mathematics as a tool for the solutions of equations was my subject of great interest after these classes. I think this is a good example of the importance of informal types of education for elementary and middle schools' students.

The next step in my trip toward applied mathematics was connected with TV programs. There were only five channels on Soviet government TV in the Moscow region in the 1970s. **Programme Three** was originally an educational channel. This channel was shown only in the major cities in the European USSR, and its programming was co-produced with the USSR Ministry of education, oriented towards the nation's student population at all levels from pre-school to college.

I was a fan of the several science programs dedicated to scientific and technical creativity on this channel, especially, "Ochevidnoye-neveroyatnoye" (Obvious and Incredible) and "This You Can". There were several special education programs for high school students on TV. These programs were short in mathematics, physics, chemistry, biology etc. Classes were taught by professors from the universities in Moscow. The programs fed students to the mathematics and physics departments in Moscow universities. It allowed college students who went through this remote TV learning to get an early start in mathematics.

Most of the professors who were involved in TV mathematics classes were from Moscow Engineering and Physics Institute (MEPhI). Naturally, I chose the (MEPhI) after graduating from high school. The MEPhI was founded in 1942 and its original mission was to train skilled personnel for the Soviet military and atomic programs. The MEPhI had a unique structure of education programs in the Soviet Union. All students had to pass a two-years general physics course, Calculus, Linear Algebra, and only after that one had a chance to choose a specific major. I chose applied mathematics as my major in third year and all students in my cohort had to take the theoretical physics courses as the students with a theoretical physics major. As a result we had good knowledge of applied math and physics. The weak point of this system was strong orientation on the applied problems (not only in nuclear physics!). As a result, we did not have interesting math courses like theory of numbers, topology etc.

Aleksei Talonov
May 31, 2021

COMMENTATORS & ANALYSTS EXTRAORDINAIRE

It has been twenty-one years since I started writing reflections, which are reincarnation of my earlier passion of writing letters. The big difference is that my reflective writings went public – from one-one to one-many, as I started sharing them with friends and relatives. And, from there on, it went to their friends and their relatives, and so on. Years ago, a student of mine created a blog for me, but seldom had I posted any reflection there. I don't have a website either. It is all emails - in a bcc mode - very old-fashioned electronically.

There are several mailing lists for communicating various categories of reflections. I have Facebook and Twitter accounts too, but they too have remained unused until a year ago. That is my approach to communication. Naturally, some readers write back and give comments. At times, a small dialog takes place. This adds clarity, expands the topic, and sharpens my thinking too.

For a number of reasons, not all the comments and commentators are included in the book– only those which are concise and strong. In a reflective style of writing, inclusion of comments adds a new flavor. Initially, I did not save readers' comments received, as publishing a book was never thought of. Also, sometimes, no comments were received. That is why the space following some reflections is blank.

Normally, an author lists the names in a book's preface or in a space for acknowledgements. I did not like that format as my readers are special. A thumbnail sketch is crafted for each one. Besides thanking them, I am sharing a piece of immortality that this book may bring! When I look at the credentials of these persons I am myself awed and wowed. These comments have come out of their incredible rich backgrounds and intimate knowledge about me and my ideas. I don't think this list can be easily matched by any other author. Here are these names in an alphabetical order.

1. Paul Aizley: 87. Emeritus professor of math at UNLV. He also served UNLV in several administrative roles. Paul is one of the few mathematicians who were also successful politicians; he served for four years as a Nevada State Assemblyman.

2. Francis A. Andrew; 66. A native of Scotland, but has been teaching English language in various programs of the Government of the Sultanate of Oman; I met him during my visiting Oman's University of Niwas in 2009. He is also an established author of a dozen books on science fiction.

3. Rahul Bhatnagar, 65: Distantly related - physician by training, works as a director of a drug safety program with a pharmaceutical company. He is an astute commentator and analyst of my ***Reflections*** - and can refine an issue to a state that it becomes indistinguishable from the one it started with.

4. HS Bhola died in May 2021 at the age of 88: Emeritus IU professor of education - friends since 1971. He often tells me how exceptional I am, as all math professors that he has known, can hardly write a sentence in English - far from being a literary writer.

5. George Buch, 50: a math lecturer at UNLV. After working as a plumber for 20 years, he went back to college, took my History of Mathematics course, and finished his master's degree in mathematics.

6. Aaron Harris, 50: instructor in the College of Southern Nevada; took three graduate courses from me, and went on to earn PhD in Mathematics Education from the College of Education.

7. Alok Kumar, 67: professor of physics at New York University at Oswego; known since 1983. He is the author of a number of books on history of science pertaining to the contribution of the Hindus

8. Sabahat Malik: 45, a native of Punjab in Pakistan, but has been teaching in the Middle East, UK and China. Met her in 2013 during my lecture at the College of Applied Sciences in Nizwa, Oman.

9. Satyam Moorty, 85: Emeritus professor of English, Southern Utah University, Cedar City. Family friends since 1976. He is a true East-west poet, essayist and enjoys his solitude.

10. Drew O'Neal, 31. My student in three graduate courses. Working for a PhD in math education.

11. JR Patadia, 75; Retired professor of MSU, Vadodara, worked as the General Secretary of Indian Mathematical Society for several years.

12. Renu Prakash; 45**;** teaches in a college, attends math meetings and follows my reflections.

13. Noel Pugache, 80: Emeritus professor of History of University of New Mexico; met him in 2012 in a Chautauqua course.

14. Shankar Raja, 85: Retired physics professor; we have been known to each other for our common interest in math conferences.

15. Viktoria Savatorova, 52. She is a native of Russia and was at UNLV for five years before moving on to the Central Connecticut State University in 2019.

16. NK Thakare, 83 is an Indian mathematician, educational administrator, writer, and a social worker. He was a vice-chancellor (equivalent to president of a US university) at a university in MS, and the General Secretary of Indian Mathematical Society.

17. Ajit Iqbal Singh; 78. Retired Math professor from Delhi University, still active in hard core math and HoM.

18. Arun Vaidya died in November 2020 at age 85, we got to know each other for our common interest in the History of Mathematics. He wrote a biography on his uncle, PC Vaidya, a well known mathematician and Gandhian.

19. Paul Werth, 53: A former Chairman of the History Department at UNLV - active in faculty governance.

20. Brian Winkel. 78: known since 1968 from the IU days. He has a unique and easy going pioneering spirit in learning a new subject, promoting it, and institutionalizing it with national and international scope.

21. Hal Whipple, 75. Known since 1974; as a lecturer, he taught math for several years, also worked for the Music Department